Lecture Notes in Mathematics
ISBN 978-3-540-08663-5 © Springer-Verlag Berlin Heidelberg

Johan L. Dupont

Curvature and Characteristic Classes

Errata

p. 5, l. 9	:	$g\omega$	should be	$g^*\omega$
p. 6, l. −5	:	Stoke's	should be	Stokes'
p. 7, l. 13	:	Stoke's	should be	Stokes'
p. 13, l. 15	:	$C^{*,k}$	should be	$C^{*-k,k}$
p. 32, l. 2	:	$(-1)^{s_1+r_1}$	should be	$(-1)^{s_1+r_2}$
p. 32, l. 3	:	$(-1)^{s_2+r_2}$	should be	$(-1)^{s_2+r_1}$
p. 51, l. 5	:	$Y_x^{g_t} - Y_x$	should be	$Y_x - Y_x^{g_t}$
p. 53, l. 15	:	$U_{\alpha \cap \beta}$	should be	$U_\alpha \cap U_\beta$
p. 54, l. 3	:	$\varphi_\alpha : U \to U_\alpha \times G$	should be	$\varphi_\alpha : \pi^{-1}U_\alpha \to U_\alpha \times G$
p. 58, l. −9	:	$X(f)\nabla_x(s)$	should be	$X(f)s$
p. 58, l. −4	:	liftet	should be	lifted
p. 64, l. −11	:	$\mathcal{A}^1(E \times [0,1])$	should be	$\mathcal{A}^1(E \times [0,1], \mathfrak{g})$
p. 67, l. 7	:	Pf	should be	$(2^{2m}\pi^m m!)Pf$
p. 74, l. 11	:	$\overline{U}_n$	should be	$\overline{U}_{n-1}$
p. 79, l. −4	:	$\pi_2\varphi_{\alpha_1}$	should be	$\pi_2 \circ \varphi_{\alpha_1}$
p. 79, l. −3	:	$V_{\alpha_0} \cap V_{\alpha_1}$	should be	$U_{\alpha_0} \cap U_{\alpha_1}$
p. 109, l. −9	:	$\psi_i \circ s \circ \varphi^{-1}$	should be	$(\varphi_i \times 1) \circ \psi_i \circ s \circ \varphi^{-1}$
p. 110, l. −8	:	$e(\xi \oplus \xi) = e(\xi) \cup e(\xi)$	should be	$e(E \oplus F) = e(E) \cup e(F)$
p. 149, l. −2 (2 places)	:	$(g_0, \ldots, g_p)$	should be	$(g_1, \ldots, g_p)$

Lecture Notes in Mathematics

Edited by A. Dold and B. Eckmann

640

Johan L. Dupont

Curvature and Characteristic Classes

Springer-Verlag
Berlin Heidelberg New York 1978

Author
Johan L. Dupont
Matematisk Institut
Ny Munkegade
DK-8000 Aarhus C/Denmark

AMS Subject Classifications (1970): 53C05, 55F40, 57D20, 58A10, 55J10

ISBN 3-540-08663-3 Springer-Verlag Berlin Heidelberg New York
ISBN 0-387-08663-3 Springer-Verlag New York Heidelberg Berlin

Printed in Germany

Printing and binding: Beltz Offsetdruck, Hemsbach/Bergstr.
2141/3140-543210

INTRODUCTION

These notes are based on a series of lectures given at the Mathematics Institute, University of Aarhus, during the academic year 1976-77.

The purpose of the lectures was to give an introduction to the classical Chern-Weil theory of characteristic classes with real coefficients presupposing only basic knowledge of differentiable manifolds and Lie groups together with elementary homology theory.

Chern-Weil theory is the proper generalization to higher dimensions of the classical Gauss-Bonnet theorem which states that for M a compact surface of genus g in 3-space

$$\frac{1}{2\pi}\int_M \kappa = 2(1-g) \tag{1}$$

where κ is the Gaussian curvature. In particular $\int_M \kappa$ is a topological invariant of M. In higher dimensions where M is a compact Riemannian manifold, $\frac{1}{2\pi}\kappa$ in (1) is replaced by a closed differential form (e.g. the Pfaffian or one of the Pontrjagin forms, see chapter 4 examples 1 and 3) associated to the curvature tensor and the integration is done over a singular chain in M. In this way there is defined a singular cohomology class (e.g. the Euler class or one of the Pontrjagin classes) which turns out to be a differential topological invariant in the sense that it depends only on the tangent bundle of M considered as a topological vector bundle.

Thus a repeating theme of this theory is to show that certain quantities which ā priori depend on the local differential geometry are actually global topological invariants. Fundamental in this context is of course the de Rham theorem which says that every real cohomology class of a manifold M can be represented by integrating a closed form over singular chains and on the other hand if integration of a closed form over singular chains represents the zero-cocycle then the form is exact. In chapter 1 we give an elementary proof of this theorem (essentially due to A. Weil [34]) which depends on 3 basic tools used several times through the lectures: (i) the integration operator of the Poincaré lemma, (ii) the nerve of a covering, (iii) the comparison theorem for double complexes (I have deliberately avoided all mentioning of spectral sequences).

In chapter 2 we show that the de Rham isomorphism respects products and for the proof we use the opportunity to introduce another basic tool: (iv) the Whitney-Thom-Sullivan theory of differential forms on simplicial sets. The resulting simplicial de Rham complex, as we call it, connects the calculus of differential forms to the combinatorial methods of algebraic topology, and one of the main purposes of these lectures is to demonstrate its applicability in the theory of characteristic classes occuring in differential geometry.

Chapter 3 contains an account of the theory of connection and curvature in a principal G-bundle (G a Lie-group) essentially following the exposition of Kobayashi and Nomizu [17]. The chapter ends with some rather long exercises (nos. 7 and 8) explaining the relation of the general theory to the classical theory of an affine connection in a Riemannian manifold.

Eventually, in chapter 4 we get to the Chern-Weil construction in the case of a principal G-bundle $\pi: E \to M$ with a connection θ and curvature Ω (in the case of a Riemannian manifold mentioned above $G = O(n)$ and E is the bundle of orthonormal tangent frames). In this situation there is associated to every G-invariant homogeneous polynomial P on the Lie algebra $\mathfrak{g}$ a closed differential form $P(\Omega^k)$ on M defining in turn a cohomology class $w_E(P) \in H^{2k}(M,\mathbb{R})$.

Before proving that this class is actually a topological invariant of the principal G-bundle we discuss in chapter 5 the general notion of a characteristic class for topological principal G-bundles. By this we mean an assignment of a cohomology class in the base space of every G-bundle such that the assignment behaves naturally with respect to bundle maps. The main theorem (5.5) of the chapter states that the ring of characteristic classes is isomorphic to the cohomology ring of the classifying space BG.

Therefore, in order to define the characteristic class $w_E(P)$ for E any topological G-bundle it suffices to make the Chern-Weil construction for the universal G-bundle EG over BG. Now the point is that although BG is not a manifold it is the realization of a simplicial manifold, that is, roughly speaking, a simplicial set where the set of p-simplices constitute a manifold. Therefore we generalize in chapter 6 the simplicial de Rham complex to simplicial manifolds, and it turns out that the Chern-Weil construction carries over to the universal bundle. In this way we get a universal Chern-Weil homomorphism

$$w\colon I^*(G) \to H^*(BG,\mathbb{R})$$

where $I^*(G)$ denotes the ring of G-invariant polynomials on the Lie algebra

In chapter 7 we specialize the construction to the classical groups obtaining in this way the Chern and Pontrjaging classes with real coefficients. We also consider the Euler class defined by the Pfaffian polynomial and in an exercise we show the Gauss-Bonnet formula in all even dimensions.

Chapter 8 is devoted to the proof of the theorem (8.1) due to H. Cartan that $w: I^*(G) \to H^*(BG,\mathbb{R})$ is an isomorphism for G a compact Lie group. At the same time we prove A.Borel's theorem that $H^*(BG,\mathbb{R})$ is isomorphic to the invariant part of $H^*(BT,\mathbb{R})$ under the Weyl group W of a maximal torus T. The corresponding result for the ring of invariant polynomials (due to C. Chevalley) depends on some Lie group theory which is rather far from the main topic of these notes, and I have therefore placed the proof in an appendix at the end of the chapter.

The final chapter 9 deals with the special properties of characteristic classes for G-bundles with a flat connection or equivalently with constant transition functions. If G is compact it follows from the above mentioned theorem 8.1 that every characteristic class with real coefficients is in the image of the Chern-Weil homomorphism and therefore must vanish. In general for $K \subseteq G$ a maximal compact subgroup we derive a formula for the characteristic classes involving integration over certain singular simplices of G/K. As an application we prove the theorem of J. Milnor [20] that the Euler number of a flat $Sl(2,\mathbb{R})$-bundle on a surface of genus h has numerical value less than h.

I have tried to make the notes as selfcontained as possible giving otherwise proper references to well-known text-books. Since our subject is classical, the literature is quite large, and especially in recent years has grown rapidly, so I have made no attempt to make the bibliography complete.

It should be noted that many of the exercises are used in the main text and also some details in the text are left as an exercise. In the course from which these notes derived the weekly exercise session played an essential role. I am grateful to the active participants in this course, especially to Johanne Lund Christiansen, Poul Klausen, Erkki Laitinen and Søren Lune Nielsen for their valuable criticism and suggestions.

Finally I would like to thank Lissi Daber for a careful typing of the manuscript and prof. Albrecht Dold and the Springer-Verlag for including the notes in this series.

Aarhus, December 15, 1977.

CONTENTS

CURVATURE AND CHARACTERISTIC CLASSES

1. Differential forms and cohomology

First let us recall the basic facts of the calculus of differential forms on a differentiable manifold M. A differential form ω of degree k associates to k C^∞ vector fields $X_1,\ldots,X_k$ a real valued C^∞ function $\omega(X_1,\ldots,X_k)$ such that it has the "tensor property" (i.e. $\omega(X_1,\ldots,X_k)_p$ depends only on $X_{1p},\ldots,X_{kp}$ for all $p \in M$) and such that it is multilinear and alternating in $X_1,\ldots,X_k$. For an l-form ω_1 and a k-form ω_2 the product $\omega_1 \wedge \omega_2$ is the $(k+l)$-form defined by

$$\omega_1 \wedge \omega_2(X_1,\ldots,X_{k+l}) =$$

$$= \frac{1}{(k+l)!}\sum_\sigma \operatorname{sign}(\sigma)\omega_1(X_{\sigma(1)},\ldots,X_{\sigma(l)})\cdot\omega_2(X_{\sigma(l+1)},\ldots,X_{\sigma(l+k)})$$

where σ runs through all permutations of $1,\ldots,k+l$. This product is associative and graded commutative, i.e.

$$\omega_1 \wedge \omega_2 = (-1)^{kl}\omega_2 \wedge \omega_1 .$$

Furthermore there is an exterior differential d which to any k-form ω associates a $(k+1)$-form $d\omega$ defined by

$$(d\omega)(X_1,\ldots,X_{k+1}) = \frac{1}{k+1}\Big[\sum_{i=1}^{k+1}(-1)^{i+1}X_i(\omega(X_1,\ldots,\hat{X}_i,\ldots,X_{k+1}))$$

$$+ \sum_{i<j}(-1)^{i+j}\omega([X_i,X_j],X_1,\ldots,\hat{X}_i,\ldots,\hat{X}_j,\ldots,X_{k+1})\Big]$$

where the "hat" means that the term is left out. Here $[X_i,X_j]$ is the Lie-bracket of the vector fields. d has the following properties:

(i) d is linear over $\mathbb{R}$

(ii) $dd = 0$

(iii) $d(\omega_1 \wedge \omega_2) = (d\omega_1) \wedge \omega_2 + (-1)^k \omega_1 \wedge d\omega_2$ for ω_1 a k-form

(iv) For a C^∞ function f and X a vector field

$(df)(X) = X(f)$

(v) d is <u>local</u>, that is, for any open set U,

$\omega | U = 0 \Rightarrow d\omega | U = 0$.

In a local coordinate system $(U, u^1, \ldots, u^n)$ any k-form ω has a unique presentation

$$\omega = \sum_{1 \leq i_1 < i_2 < \ldots < i_k \leq n} a_{i_1, \ldots, i_k} du^{i_1} \wedge du^{i_2} \wedge \ldots \wedge du^{i_k}$$

where $a_{i_1 \ldots i_k}$ are C^∞ functions on U.

Suppose $F : M \to N$ is a C^∞ map of C^∞ manifolds and let ω be a k-form on N. Then there is a unique <u>induced</u> k-form $F^*\omega$ on M such that for any k vector fields $X_1, \ldots, X_k$ on M

$$F^*(\omega)(X_1, \ldots, X_k)_q = \omega_{F(q)}(F_* X_{1q}, \ldots, F_* X_{kq}), \quad \forall q \in M,$$

where $F_* = dF$ is the differential of F. F^* preserves $\wedge$ and commutes with d.

The set of k-forms on M is denoted $A^k(M)$ and we shall refer to $(A^*(M), \wedge, d)$ or just $A^*(M)$ as the <u>de Rham complex</u> (or <u>de Rham algebra</u>) of M.

For $U \subseteq M$ an open subset $A^*(U)$ is clearly defined since U is a manifold. Now suppose $U \subseteq M$ is a <u>closed</u> subset of M and suppose that every point of U is a limit of interior points of U. Then at any point $q \in U$ the tangent space $T_q(U)$ is naturally identified with $T_q(M)$. By a k-form on U we shall understand a collection ω_q of k-linear alternating forms on

$T_q(M)$, $q \in U$, which extends to a differential form on all of M (it is enough that it extends to an open neighbourhood of U by a "bump function" argument). Again let $A^k(U)$ denote the set of k-forms on U. Notice that a differential form on U is determined by its restriction to the interior of U. Therefore $d : A^k(U) \to A^{k+1}(U)$ is well-defined and we again have a de Rham complex $(A^*(U), \wedge, d)$. This observation is important because of the following example:

<u>Example 1</u>. <u>The standard n-simplex Δ^n</u>. In $\mathbb{R}^{n+1}$ consider Δ^n, the convex hull of the set of canonical basis vectors $e_i = (0,0,\dots,1,0,\dots,0)$ with 1 on the i-th place, $i = 0,1,\dots,n$. That is

$$\Delta^n = \{t = (t_0,\dots,t_n) \mid t_i \geqq 0,\ i = 0,\dots,n,\ \textstyle\sum_j t_j = 1\}$$

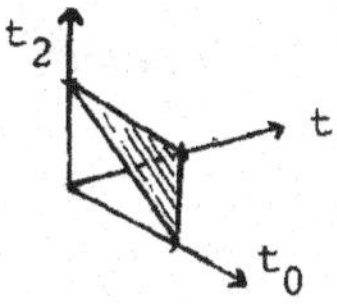

The hyperplane $V^n = \{t \in \mathbb{R}^{n+1} \mid \sum_j t_j = 1\}$ is clearly a manifold and $\Delta^n \subseteq V^n$ is clearly the closure of its interior points in V^n. So it makes sense to talk about $A^k(\Delta^n)$. Considering the barycentric coordinates $(t_0,\dots,t_n)$ as functions on V^n we have their differentials dt_i, $i = 0,\dots,n$, and every k-form on V^n (or Δ^n) is expressible in the form

$\sum_{0 \leqq i_0 < \dots < i_k \leqq n} a_{i_0 \dots i_k}\, dt_{i_0} \wedge \dots \wedge dt_{i_k}$ where $a_{i_0 \dots i_k}$ are C^∞ functions on V^n (or Δ^n). Notice that the relation $t_0 + \dots + t_n = 1$ implies $dt_0 + \dots + dt_n = 0$, so actually the set $\{dt_1,\dots,dt_n\}$ generates $A^*(\Delta^n)$.

Now return to $U \subseteq M$ an open or closed subset of a C^∞ manifold as before. A k-form ω on U is called <u>closed</u> if $d\omega = 0$ and ω is <u>exact</u> if $\omega = d\omega'$ for some $(k-1)$-form ω'. Since $dd = 0$ every exact form is closed.

<u>Definition 1.1</u>. The k-th <u>de Rham cohomology group</u> of U is the real vectorspace

$$H^k(A^*(U)) = \ker(d : A^k(U) \to A^{k+1}(U))/dA^{k-1}(U)$$
$$k = 0,1,2,\ldots \quad (A^{-1}(U) = 0).$$

<u>Example 2</u>. For $M = \mathbb{R}^2$ with coordinates (x,y) any 1-form is of the form $\omega = fdx + gdy$ and $d\omega = 0$ is just the requirement

$$\frac{\partial f}{\partial y} = \frac{\partial g}{\partial x}.$$

Now take $U = \mathbb{R}^2 \setminus \{0\}$ and consider the 1-form

$$\omega = \frac{1}{x^2+y^2}(xdy-ydx).$$

It is easily seen that ω is closed but $\int_{S^1} \omega = 2\pi$ so ω is not exact. Hence $H^1(A^*(U)) \neq 0$.

It is classically wellknown that $H^*(A^*(M))$ is related to the geometry of M. For example let $U \subseteq \mathbb{R}^n$ be <u>star-shaped</u> with respect to $e \in U$, that is, for all $x \in U$ the whole line segment from e to x is contained in U. Then we have:

<u>Lemma 1.2</u>. (Poincaré's lemma). Let $U \subseteq \mathbb{R}^n$ be star-shaped with respect to $e \in U$. Then there are operators $h^k : A^k(U) \to A^{k-1}(U)$, $k = 1,2,\ldots,$ such that for any $\omega \in A^k(U)$,

$$(1.3) \qquad h^{k+1}(d\omega) = \begin{cases} -\omega - dh^k(\omega), & k \gtrless 0 \\ \omega(e) - \omega, & k = 0. \end{cases}$$

In particular

$$(1.4) \qquad H^k(A^*(U)) = \begin{cases} 0, & k > 0 \\ \mathbb{R}, & k = 0. \end{cases}$$

Proof. Clearly (1.4) follows from (1.3).

The operators h^k are defined as follows: let $g : [0,1] \times U \to U$ be the map

$$g(s,x) = se + (1-s)x, \quad s \in [0,1], \quad x \in U.$$

For any $\omega \in A^k(U)$, $g^*\omega \in A^k([0,1] \times U)$ is uniquely expressible as

$$g\ \omega = ds \wedge \alpha + \beta$$

where α and β are forms not involving ds. (The $(k-1)$-form α is usually denoted $i_{\frac{\partial}{\partial s}}(g^*\omega)$.) Then define

$$h^k(\omega) = \int_{s=0}^{1} \alpha$$

which means that we integrate the coefficients of α with respect to the variable s. In order to prove (1.3) notice that

$$g^*d\omega = d(g^*\omega) = -ds \wedge d_x\alpha + ds \wedge \frac{\partial}{\partial s}\beta + \ldots$$

where we have only written the terms involving ds, and where $d_x\alpha = d\alpha - ds \wedge \frac{\partial}{\partial s}\alpha$. Hence

$$h^{k+1}(d\omega) = \int_{s=0}^{1} \frac{\partial}{\partial s}\beta - d_x\alpha.$$

For $k = 0$ clearly $\alpha = 0$ so

$$h^1(d\omega)(x) = \int_{s=0}^{1} \frac{\partial}{\partial s}\ \omega(se+(1-s)x) = \omega(e) - \omega(x), \quad x \in U.$$

For $k > 0$,

$$\beta | 0 \times U = (id)^*\omega = \omega$$

$$\beta | 1 \times U = g_1^*\omega = 0, \qquad g_1(x) = e, \quad x \in U.$$

Hence

$$h^{k+1}(d\omega) = -\omega - d\int_{s=0}^{1} \alpha = -\omega - dh^k(\omega),$$

which proves (1.3).

The de Rham theorem which is the main object of this chapter gives a geometric interpretation of the de Rham cohomology of a general manifold. First we need a few remarks about integration of forms. Actually we shall only integrate n-forms over the standard n-simplex Δ^n. The orientation on Δ^n or rather V^n is determined by the n-form $dt_1 \wedge \ldots \wedge dt_n$. Explicitly every n-form on Δ^n is uniquely expressible as

$$\omega = f(t_1,\ldots,t_n)dt_1 \wedge \ldots \wedge dt_n$$

and by definition

$$\int_{\Delta^n} \omega = \int_{\Delta^n} f(t_1,\ldots,t_n)dt_1\ldots dt_n$$

where $\Delta^n \subseteq \mathbb{R}^n$ is the set $\Delta_0^n = \{(t_1,\ldots,t_n) \in \mathbb{R}^n |\ t_i \geqq 0, \sum_j t_j \leqq 1\}$,

<u>Exercise 1</u>. Show that

$$(1.5) \qquad \int_{\Delta^n} dt_1 \wedge \ldots \wedge dt_n = \frac{1}{n!}\ .$$

<u>Exercise 2</u>. Show the following case of <u>Stoke's theorem</u>: Let $\varepsilon^i : \Delta^{n-1} \to \Delta^n$, $i = 0,\ldots,n$, be the <u>face map</u>

$$\varepsilon^i(t_0,\ldots,t_{n-1}) = (t_0,\ldots,t_{i-1},0,t_i,\ldots,t_{n-1}).$$

Let $\omega \in A^{n-1}(\Delta^n)$. Then

$$(1.6) \qquad \int_{\Delta^n} d\omega = \sum_{i=0}^{n} (-1)^i \int_{\Delta^{n-1}} (\varepsilon^i)^*\omega.$$

(Hint: First show a similar formula for the cube $I^n \subset \mathbb{R}^n$, $I = [0,1]$, (see e.g. M. Spivak [29, p. 8-18], Then deduce (1.6) by using the map $g : I^n \to \Delta^n$ given by

$$g(s_1,\dots,s_n) = ((1-s_1), s_1(1-s_2), s_1 s_2(1-s_3), \dots$$

$$\dots, s_1 \dots s_{n-1}(1-s_n), s_1 \dots s_n).)$$

<u>Exercise 3</u>. Δ^n is clearly star-shaped with respect to each of the vertices e_i, $i = 0,1,\dots,n$. By lemma 1.2 we therefore have $n + 1$ corresponding operators $h_{(i)}$: $A^k(\Delta^n) \to A^{k-1}(\Delta^n)$, $k = 1,2,\dots,$ satisfying (1.3) with $e = e_i$, $i = 0,\dots,n$. Show that for any n-form ω on Δ^n

$$(1.7) \qquad \int_{\Delta^n} \omega = (-1)^n h_{(n-1)} \circ \dots \circ h_{(0)}(\omega)(e_n).$$

(Hint: First show that the operator on the right satisfies Stoke's theorem (equation (1.6) above) and then use induction.)

Now let us recall the elementary facts about singular homology and cohomology. We consider the case of C^∞ manifolds and C^∞ maps which is completely analogous to the case of topological spaces and continuous maps usually considered. Also we shall only use the field of real numbers $\mathbb{R}$ as coefficient ring.

Let M be a C^∞ manifold. A C^∞ <u>singular</u> <u>n-simplex</u> in M is a C^∞ map $\sigma : \Delta^n \to M$, where Δ^n is the standard n-simplex. Let $S_n^\infty(M)$ denote the set of all C^∞ singular n-simplices in M. As in exercise 2 above let $\varepsilon^i : \Delta^{n-1} \to \Delta^n$, $i = 0,\dots,n$, be the inclusion on the i-th face. Define $\varepsilon_i : S_n^\infty(M) \to S_{n-1}^\infty(M)$, $i = 0,\dots,n$, by

$$\varepsilon_i(\sigma) = \sigma \circ \varepsilon^i, \qquad \sigma \in S_n^\infty(M).$$

Notice that

(1.8) $\qquad \varepsilon_i \circ \varepsilon_j = \varepsilon_{j-1} \circ \varepsilon_i \quad \text{if} \quad i < j.$

The group of C^∞ <u>singular</u> n-chains with coefficients in $\mathbb{R}$ is the free vector space $C_n(M)$ on $S_n^\infty(M)$, i.e. the vector space of finite formal sums $\sum_{\sigma \in S_n^\infty(M)} a_\sigma \cdot \sigma$. The maps ε_i, $i = 0,\ldots,n$, clearly extend to $\varepsilon_i : C_n(M) \to C_{n-1}(M)$ and we have the <u>boundary operator</u> $\partial = \sum_i (-1)^i \varepsilon_i : C_n(M) \to C_{n-1}(M)$. (1.8) implies that $\partial\partial = 0$ and we have the n-th C^∞ <u>singular homology group</u> with real coefficients

$$H_n(M) = H_n(C_*(M)) = \ker(\partial : C_n(M) \to C_{n-1}(M))/\partial C_{n-1}(M).$$

Dually the group of C^∞ <u>singular n-cochains</u> with real coefficients is

$$C^n(M) = \mathrm{Hom}(C_n(M), \mathbb{R})$$

and we have the <u>coboundary</u> $\delta = \partial^* : C^n(M) \to C^{n+1}(M)$. Explicitly an n-cochain is a function $c : S_n^\infty(M) \to \mathbb{R}$ or equivalently a collection $c = \{c_\sigma\}$, $\sigma \in S_n^\infty(M)$, of real numbers, and δ is given by

(1.9) $$(\delta c)_\tau = \sum_{i=0}^{n+1} (-1)^i c_{\varepsilon_i \tau}, \qquad \tau \in S_{n+1}^\infty(M).$$

Again the n-th C^∞ singular <u>cohomology</u> group with real coefficients is

$$H^n(M) = H^n(C^*(M)) = \ker(\delta : C^n(M) \to C^{n+1}(M))/\delta C^{n-1}(M).$$

If $f : M \to N$ is a C^∞ map of C^∞ manifolds we clearly get an induced map $S(f) : S_*^\infty(M) \to S_*^\infty(N)$ defined by $S(f)(\sigma) = f \circ \sigma$. This clearly extends to $f_\# : C_*(M) \to C_*(N)$ and dually induces $f^\# : C^*(N) \to C^*(M)$. Obviously C_* and C^* are covariant and contravariant functors respectively. Also $f_\#$ and $f^\#$ are chain-maps, i.e.

$$f_{\#} \circ \partial = \partial \circ f_{\#}, \quad \delta \circ f^{\#} = f^{\#} \circ \delta.$$

Therefore we have induced maps

$$f_* : H_*(M) \to H_*(N), \quad f^* : H^*(N) \to H^*(M).$$

Let us recall the following wellknown facts:

(1.10) $H_i(pt) = 0, \quad i > 0, \quad H_0(pt) = \mathbb{R}$

$H^i(pt) = 0, \quad i > 0, \quad H^0(pt) = \mathbb{R}.$

(1.11) (Homotopy property). Suppose $f_0, f_1 : M \to N$ are homotopic, i.e., there is a C^∞ map $F : M \times [0,1] \to N$ such that $F|M \times 0 = f_0$, $F|M \times 1 = f_1$. Then $f_{0\#}$ and $f_{1\#}$ are chain homotopic, i.e., there are homomorphisms $s_i : C_i(M) \to C_{i+1}(N)$ such that

$$f_{1\#} - f_{0\#} = s_{i-1} \circ \partial + \partial \circ s_i.$$

In particular

$$f_{1*} = f_{0*} : H_*(M) \to H_*(N),$$

$$f_1^* = f_0^* : H^*(N) \to H^*(M).$$

(1.12) (Excision property). Suppose $\mathcal{U} = \{U_\alpha\}_{\alpha \in \Sigma}$ is an open covering of M and let $S_n^\infty(\mathcal{U})$ denote the set of singular n-simplices of M, $\sigma : \Delta^n \to M$, such that $\sigma(\Delta^n) \subsetneq U_\alpha$ for some α. Let $(C^*(\mathcal{U}), \partial)$ and $(C^*(\mathcal{U}), \delta)$ be the corresponding chain or cochain complexes (called "with support in $\mathcal{U}$") and let

$$\iota_* : C_*(\mathcal{U}) \to C_*(M), \quad \iota^* : C^*(M) \to C^*(\mathcal{U})$$

be the natural maps induced by the inclusion

$\iota : S_n^\infty(\mathcal{U}) \subset S_n^\infty(M)$. Then ι_* and ι^* are <u>chain equivalences</u>, in particular they induce isomorphisms

$$H(C_*(\mathcal{U})) \cong H(C_*(M)), \quad H(C^*(M)) \cong H(C^*(\mathcal{U})).$$

We now define a natural map

$$I : A^n(M) \to C^n(M)$$

by the formula

$$(1.13) \qquad I(\omega)_\sigma = \int_{\Delta^n} \sigma^*\omega, \qquad \omega \in A^n(M), \quad \sigma \in S_n^\infty(M).$$

I is clearly a <u>natural transformation</u> of functors, that is, if $f : M \to N$ is a C^∞ map, then

$$I \circ f^* = f^\# \circ I,$$

where $f^* : A^*(N) \to A^*(M)$ and $f^\# : C^*(N) \to C^*(M)$ are the induced maps.

<u>Lemma 1.14</u>. I is a chain map, i.e.

$$I \circ d = \delta \circ I.$$

In particular I induces a map on homology

$$I : H(A^*(M)) \to H(C^*(M)).$$

<u>Proof</u>. This simply follows using exercise 2 above:

$$\begin{aligned} I(d\omega)_\tau &= \int_{\Delta^{n+1}} \tau^*(d\omega) = \int_{\Delta^{n+1}} d\tau^*\omega \\ &= \sum_{i=0}^{n+1} (-1)^i \int_{\Delta^n} (\varepsilon^i)^*\tau^*\omega = \sum_{i=0}^{n+1} (-1)^i \int_{\Delta^n} (\varepsilon_i(\tau))^*\omega \\ &= \sum_{i=0}^{n+1} (-1)^i\, I(\omega)_{\varepsilon_i(\tau)} = \delta(I(\omega))_\tau, \end{aligned}$$

$$\omega \in A^n(M), \quad \tau \in S_{n+1}^\infty(M).$$

Theorem 1.15. (de Rham). $I : H^*(A^*(M)) \to H(C^*(M))$ is an isomorphism for any C^∞ manifold M.

First notice:

Lemma 1.16. Theorem 1.15 is true for M diffeomorphic to a star shaped open set in $\mathbb{R}^n$.

Proof. It is clearly enough to consider $M = U \subseteq \mathbb{R}^n$ an open set star shaped with respect to $e \in U$. As in Lemma 1.2 consider the homotopy $g : U \times [0,1] \to U$ with $g(-,1) = id$ and $g(-,0) = e$ given by

$$g(x,s) = sx + (1-s)e.$$

By (1.11) the inclusion $\{e\} \subsetneq U$ induces an isomorphism in singular cohomology, so the statement follows from (1.10) together with Lemma 1.2 and the commutative diagram

$$\begin{array}{ccc} H(A^*(U)) & \xrightarrow{\quad I \quad} & H(C^*(U)) \\ \cong\downarrow & & \downarrow\cong \\ H(A^*(e)) & \xrightarrow{\quad I \quad} & H(C^*(e)) \\ \| & & \| \\ \mathbb{R} & \longrightarrow & \mathbb{R} \end{array}$$

Lemma 1.17. For any C^∞ manifold M of dimension n there is an open covering $\mathcal{U} = \{U_\alpha\}_{\alpha \in \Sigma}$, such that every non-empty finite intersection $U_{\alpha_0} \cap \ldots \cap U_{\alpha_p}$, $\alpha_0, \ldots, \alpha_p \in \Sigma$, is diffeomorphic to a star shaped open set of $\mathbb{R}^n$.

Proof. Choose a Riemannian metric on M. Then every point has a neighbourhood U which is normal with respect to every point of U (i.e., for every $q \in U$, $\exp_q$ is a diffeomorphism of a star shaped neighbourhood of $0 \in T_q(M)$ onto U). In particular, U is geodesically convex, that is, for every pair of points $p,q \in U$ there is a unique geodesic segment in M joining p and q and this is contained in U. (For a proof see e.g. S. Helgason [14, Chapter I Lemma 6.4). Now choose a covering $\mathcal{U} = \{U_\alpha\}_{\alpha\in\Sigma}$ with such open sets. Then any non-empty finite intersection $U_{\alpha_0} \cap \ldots \cap U_{\alpha_k}$ is again geodesically convex and so is a normal neighbourhood of each of its points. It is therefore clearly diffeomorphic to a star shaped region in $\mathbb{R}^n$ (via the exponential map).

In view of the last two lemmas it is obvious that we want to prove Theorem 1.15 by some kind of formal inductive argument using a covering as in Lemma 1.17. What is needed are some algebraic facts about <u>double complexes</u>:

We consider modules over a fixed ring R (actually we shall only use $R = \mathbb{R}$). A <u>complex</u> C^* is a $\mathbb{Z}$-graded module with a differential $d : C^n \to C^{n+1}$, $n \in \mathbb{Z}$, such that $dd = 0$. Similarly, a <u>double complex</u> is a $\mathbb{Z} \times \mathbb{Z}$-graded module $C^{*,*} = \coprod_{p,q} C^{p,q}$, together with two differentials

$$d' : C^{p,q} \to C^{p+1,q}, \quad d'' : C^{p,q} \to C^{p,q+1}$$

satisfying

$$(1.18) \qquad d'd' = 0, \quad d''d'' = 0, \quad d''d' + d'd'' = 0.$$

We shall actually assume that $C^{*,*}$ is a 1. quadrant double complex, that is, $C^{p,q} = 0$ if either $p < 0$ or $q < 0$. Associated to $(C^{*,*},d',d'')$ is the <u>total complex</u> (C^*,d) where

$$C^n = \coprod_{p+q=n} C^{p,q}, \quad d = d' + d''.$$

For fixed q we can take the homology of $C^{*,q}$ with respect to d'. This gives another bi-graded module $E_1^{p,q} = H^p(C^{*,q},d')$.

Now suppose ${}_1C^{*,*}$ and ${}_2C^{*,*}$ are two double complexes as above, and suppose $f : {}_1C^{*,*} \to {}_2C^{*,*}$ is a homomorphism respecting the grading and commuting with d' and d''. Then clearly f gives a chain map of the associated total complexes and hence induces $f_* : H({}_1C^*,d) \to H({}_2C^*,d)$. Also clearly f induces $f_1 : {}_1E_1^{p,q} \to {}_2E_1^{p,q}$. We now have:

Lemma 1.19. Suppose $f : {}_1C^{*,*} \to {}_2C^{*,*}$ is a homomorphism of 1. quadrant double complexes and suppose $f_1 : {}_1E_1^{*,*} \to {}_2E_1^{*,*}$ is an isomorphism. Then also $f_* : H({}_1C^*) \to H({}_2C^*)$ is an isomorphism.

Proof. For a double complex $(C^{*,*},d',d'')$ with total complex (C^*,d) define the subcomplexes $F_q^* \subseteq C^*$, $q \in \mathbb{Z}$, by

$$F_q^* = \coprod_{k \geq q} C^{*,k}.$$

$k+q = *$

Then clearly

$$\cdots \supseteq F_{q-1}^* \supseteq F_q^* \supseteq F_{q+1}^* \supseteq \cdots$$

and $d : F_q^* \to F_q^*$. Notice that the complex $(F_q^*/F_{q+1}^*,d)$ is isomorphic to $(C^{*,q},d')$. Therefore for $f : {}_1C^{*,*} \to {}_2C^{*,*}$ a map of double complexes the assumption that $f_1 : {}_1E_1^{p,q} \to {}_2E_1^{p,q}$ is an isomorphism, is equivalent to saying that $f : {}_1F_q^*/{}_1F_{q+1}^* \to {}_2F_q^*/{}_2F_{q+1}^*$, $q \in \mathbb{Z}$, induces an isomorphism in homology. Now by induction for $r = 1,2,\ldots$ it follows from the commutative diagram of chain complexes

$$\begin{array}{ccccccccc} 0 & \to & {}_1F_{q+r}^*/{}_1F_{q+r+1}^* & \to & {}_1F_q^*/{}_1F_{q+r+1}^* & \to & {}_1F_q^*/{}_1F_{q+r}^* & \to & 0 \\ & & \downarrow f & & \downarrow f & & \downarrow f & & \\ 0 & \to & {}_2F_{q+r}/{}_2F_{q+r+1}^* & \to & {}_2F_q^*/{}_2F_{q+r+1}^* & \to & {}_2F_q^*/{}_2F_{q+r}^* & \to & 0 \end{array}$$

and the five lemma that

$$f : {}_1F_q^*/{}_1F_{q+r}^* \to {}_2F_q^*/{}_2F_{q+r}^*$$

induces an isomorphism in homology for all $q \in \mathbb{Z}$ and $r = 1,2,\ldots$. However, for a 1. quadrant double complex $C^{*,*}$ we have

$$F_0^n = C^n \quad \text{and} \quad F_r^n = 0 \quad \text{for} \quad r > n$$

so the lemma follows.

Remark. Interchanging p and q in $C^{p,q}$ we get a similar lemma with $E_1^{p,q}$ replaced by $H^q(C^{p,*},d")$.

Notice that for a 1. quadrant double complex $C^{*,*}$ it follows from (1.18) that $d"$ induces a differential also denoted $d" : E_1^{p,q} \to E_1^{p,q+1}$ for each p. In particular, since

$$E_1^{0,q} = \ker(d' : C^{0,q} \to C^{1,q}) \subseteq C^{0,q} \subseteq C^q$$

we have a natural inclusion of chain complexes

$$e : (E_1^{0,*},d") \to (C^*,d)$$

(called the "edge-homomorphism").

Corollary 1.20. Suppose $E_1^{p,q} = 0$ for $p > 0$. Then e induces an isomorphism

$$e_* : H(E_1^{0,*},d") \to H(C^*,d).$$

Proof. $E_1^{p,q}$ is a double complex with $d' = 0$. Apply Lemma 1.19 for the natural inclusion $E_1^{p,q} \to C^{p,q}$.

Note. For more information on double complexes see e.g. G. Bredon [7, appendix] or S. Mac Lane [18, Chapter 11, §§ 3 and 6].

We now turn to

Proof of Theorem 1.15. Choose a covering $\mathcal{U} = \{U_\alpha\}_{\alpha\in\Sigma}$ of M as in Lemma 1.17. Associated to this we get a double complex as follows: Given $p,q \geq 0$ consider

$$A_{\mathcal{U}}^{p,q} = \prod_{(\alpha_0,\dots,\alpha_p)} A^q(U_{\alpha_0} \cap\dots\cap U_{\alpha_p})$$

where the product is over all ordered (p+1)-tuples $(\alpha_0,\dots,\alpha_p)$ with $\alpha_i \in \Sigma$ such that $U_{\alpha_0}\cap\dots\cap U_{\alpha_p} \neq \emptyset$. The "vertical" differential is given by

$$(-1)^p d : A_{\mathcal{U}}^{p,q} \to A_{\mathcal{U}}^{p,q+1},$$

where $d : A^q(U_{\alpha_0} \cap\dots\cap U_{\alpha_p}) \to A^{q+1}(U_{\alpha_0} \cap\dots\cap U_{\alpha_p})$ is the exterior differential operator. The "horizontal" differential

$$\delta : A_{\mathcal{U}}^{p,q} \to A_{\mathcal{U}}^{p+1,q}$$

is given as follows:

For $\omega = (\omega_{(\alpha_0,\dots,\alpha_p)}) \in A_{\mathcal{U}}^{p,q}$ the component of $\delta\omega$ in $A^q(U_{\alpha_0} \cap\dots\cap U_{\alpha_{p+1}})$ is given by

$$(1.21) \qquad (\delta\omega)_{(\alpha_0,\dots,\alpha_{p+1})} = \sum_{i=0}^{p+1} (-1)^i \omega_{(\alpha_0,\dots,\hat{\alpha}_i,\dots,\alpha_{p+1})}.$$

It is easily seen that $\delta\delta = 0$ and $\delta d = d\delta$ so $A_{\mathcal{U}}^{p,q}$ is a double complex.

Now notice that there is a natural inclusion

$$A^q(M) \hookrightarrow \prod_{\alpha_0} A^q(U_{\alpha_0}) = A_{\mathcal{U}}^{0,q}.$$

Lemma 1.22. For each q the sequence

$$0 \to A^q(M) \to A_{\mathcal{U}}^{0,q} \to A_{\mathcal{U}}^{1,q} \to \ldots$$

is exact.

Proof. In fact putting $A_{\mathcal{U}}^{-1,q} = A^q(M)$ we can construct homomorphisms

$$s_p : A_{\mathcal{U}}^{p,q} \to A_{\mathcal{U}}^{p-1,q}$$

such that

(1.23) $$s_{p+1} \circ \delta + \delta \circ s_p = \mathrm{id}.$$

To do this just choose a partition of unity $\{\varphi_\alpha\}_{\alpha\in\Sigma}$ with $\operatorname{supp}\varphi_\alpha \subseteq U_\alpha$, $\forall\alpha \in \Sigma$, and define

$$(s_p\omega)(\alpha_0,\ldots,\alpha_{p-1}) = (-1)^p \sum_{\alpha\in\Sigma} \varphi_\alpha \omega(\alpha_0,\ldots,\alpha_{p-1},\alpha),$$
$$\omega \in A_{\mathcal{U}}^{p,q}.$$

It is easy to verify that s_p is well-defined and that (1.23) is satisfied.

It follows that

$$E_1^{p,q} = \begin{cases} 0, & p > 0 \\ A^q(M), & p = 0. \end{cases}$$

Together with Corollary 1.20 this proves

Lemma 1.24. Let $A_{\mathcal{U}}^*$ be the total complex of $A_{\mathcal{U}}^{*,*}$. Then there is a natural chain map

$$e_A : A^*(M) \to A_{\mathcal{U}}^*$$

which induces an isomorphism in homology.

We now want to do the same thing with A^* replaced by the singular cochain functor C^*. As before we get a double complex

$$C_{\mathcal{U}}^{p,q} = \prod_{(\alpha_0,\ldots,\alpha_p)} C^q(U_{\alpha_0} \cap \ldots \cap U_{\alpha_p})$$

where the "vertical" differential is given by $(-1)^p$ times the coboundary in the complex $C^*(U_{\alpha_0} \cap \ldots \cap U_{\alpha_p})$ and where the "horizontal" differential is given by the same formula as (1.21) above. Again we have a natural map of chain complexes

$$e_C : C^*(M) \to C_{\mathcal{U}}^{0,*} \subseteq C_{\mathcal{U}}^*$$

and we want to prove

Lemma 1.25. $e_C : C^*(M) \to C_{\mathcal{U}}^*$ induces an isomorphism in homology.

Suppose for the moment that Lemma 1.25 is true and let us finish the proof of Theorem 1.15 using this.

For $U \subseteq M$ we have a chain map

$$I : A^*(U) \to C^*(U)$$

as defined by (1.13) above. Therefore we clearly get a map of double complexes

$$I : A_{\mathcal{U}}^{p,q} \to C_{\mathcal{U}}^{p,q}$$

and we have a commutative diagram

$$\begin{array}{ccc} A_{\mathcal{U}}^* & \longrightarrow & C_{\mathcal{U}}^* \\ \uparrow e_A & & \uparrow e_C \\ A^*(M) & \longrightarrow & C^*(M) \end{array}$$

By (1.24) and (1.25) the vertical maps induce isomorphisms in homology. It remains to show that the upper horizontal map induces an isomorphism in homology. Now by the remark following Lemma 1.19 it suffices to see that for each p

$$I : H(A_{\mathcal{U}}^{p,*}) \to H(C_{\mathcal{U}}^{p,*})$$

is an isomorphism. However this is exactly Lemma 1.16 applied to each of the sets $U_{\alpha_0} \cap \ldots \cap U_{\alpha_p}$.

Proof of Lemma 1.25. It is not true that Lemma 1.22 holds with A^* replaced by C^*. However, if we restrict to cochains with support in the covering $\mathcal{U}$ it is true. Thus as in (1.12) let $C^q(\mathcal{U})$ denote the q-cochains defined on simplices $\sigma \in S_q^\infty(\mathcal{U})$, i.e. for each $\sigma \in S_q^\infty(\mathcal{U})$ there is a U_α with $\sigma(\Delta^q) \subseteq U_\alpha$. Then there is a natural restriction map $C^q(\mathcal{U}) \to C_{\mathcal{U}}^{0,q}$ and the sequence

(1.26) $$0 \to C^q(\mathcal{U}) \to C_{\mathcal{U}}^{0,q} \to C_{\mathcal{U}}^{1,q} \to \ldots$$

is exact. In fact we construct homomoprhisms

$$s_p : C_{\mathcal{U}}^{p,q} \to C_{\mathcal{U}}^{p-1} \qquad (C_{\mathcal{U}}^{-1,q} = C^q(\mathcal{U})),$$

as follows: For each $\sigma \in S_q^\infty(\mathcal{U})$ choose $\alpha(\sigma) \in \sum$ such that $\sigma(\Delta^q) \subseteq U_{\alpha(\sigma)}$, and define

$$s_p(c)_{(\alpha_0,\ldots,\alpha_{p-1})}(\sigma) = (-1)^p c_{(\alpha_0,\ldots,\alpha_{p-1},\alpha(\sigma))}(\sigma).$$

Then an easy calculation shows that

$$s_{p+1} \circ \delta + \delta \circ s_p = \mathrm{id}.$$

It follows that the chain map $e_C : C^*(M) \to C_{\mathcal{U}}^*$ factors into $e_C = \bar{e}_C \circ \iota^*$, where $\iota^* : C^*(M) \to C^*(\mathcal{U})$ is the natural chain map as in (1.12) and where the edge homomorphism

$$\bar{e}_C : C^*(\mathcal{U}) \to C^*_{\mathcal{U}}$$

induces an isomorphism in homology by Corollary 1.20 and the exactness of (1.26). Since ι^* also induces an isomorphism in homology by (1.12) this ends the proof of Lemma 1.25 and also of Theorem 1.15.

Exercise 4. For a topological space X let $S_n^{top}(X)$ denote the set of continuous singular n-simplices of X, and let $C_*^{top}(X)$ and $C^*_{top}(X)$ be the corresponding chain and cochain complexes. Show that for a C^∞ manifold M the inclusion $S_*^\infty(M) \to S_*^{top}(M)$ induces isomorphisms in homology

$$H(C_*(M)) \to H(C_*^{top}(M)), \quad H(C^*_{top}(M)) \to H(C^*(M)).$$

(Hint: Use double complexes for a covering as in Lemma 1.17).

Hence the homology and cohomology based on C^∞ singular simplices agree with the usual singular homology and cohomology. It follows therefore from Theorem 1.15 that the de Rham cohomology groups are topological invariants.

Exercise 5. Show directly the analogue of the homotopy property (1.11) for the de Rham complex.

Note. The above proof of de Rham's theorem goes back to A. Weil [34]. It contains the germs of the theory of sheaves. For an exposition of de Rham's theorem in this context see e.g. F. W. Warner [33, chapter 5].

2. Multiplicativity. The simplicial de Rham complex

In Chapter 1 we showed that for a differentiable manifold M the de Rham cohomology groups $H^k(A^*(M))$ are topological invariants of M. As mentioned above the wedge-product

$$\wedge : A^k(M) \otimes A^l(M) \to A^{k+l}(M) \tag{2.1}$$

makes $A^*(M)$ an algebra and it is easy to see that (2.1) induces a multiplication

$$\wedge : H^k(A^*(M)) \otimes H^l(A^*(M)) \to H^{k+l}(A^*(M)). \tag{2.2}$$

In this chapter we shall show that (2.2) is also a topological invariant. More precisely, let

$$\smile : H^k(C^*(M)) \otimes H^l(C^*(M)) \to H^{k+l}(C^*(M)) \tag{2.3}$$

be the usual cup-product in singular cohomology; then we shall prove

<u>Theorem 2.4</u>. For any differentiable manifold M the diagram

$$\begin{array}{ccc} H^k(A^*(M)) \otimes H^l(A^*(M)) & \xrightarrow{\quad\wedge\quad} & H^{k+l}(A^*(M)) \\ \downarrow I \otimes I & & \downarrow I \\ H^k(C^*(M)) \otimes H^l(C^*(M)) & \xrightarrow{\quad\smile\quad} & H^{k+l}(C^*(M)) \end{array}$$

commutes.

For the proof it is convenient to introduce the <u>simplicial de Rham complex</u> which is a purely combinatorial construction closely related to the cochain complex C^* but on the other hand has the same formal properties as the de Rham complex A^*.

We shall define it for a general simplicial set:

Definition 2.5. A simplicial set S is a sequence $S = \{S_q\}$, $q = 0,1,2,\ldots,$ of sets together with face operators $\varepsilon_i : S_q \to S_{q-1}$, $i = 0,\ldots,q$, and degeneracy operators $\eta_i : S_q \to S_{q+1}$, $i = 0,\ldots,q$, which satisfy the identities

$$\text{(i)} \qquad \varepsilon_i\varepsilon_j = \varepsilon_{j-1}\varepsilon_i, \qquad i < j,$$

$$\text{(ii)} \qquad \eta_i\eta_j = \eta_{j+1}\eta_i, \qquad i \leqq j,$$

$$\text{(iii)} \qquad \varepsilon_i\eta_j = \begin{cases} \eta_{j-1}\varepsilon_i, & i < j, \\ id, & i = j,\ i = j+1, \\ \eta_j\varepsilon_{i-1}, & i > j+1. \end{cases}$$

Example 1. We shall mainly consider the example, where $S_q = S_q^\infty(M)$ or $S_q^{top}(M)$. Here as in Chapter 1, $\varepsilon_i(\sigma) = \sigma \circ \varepsilon^i$, $i = 0,\ldots,q$, where $\varepsilon^i : \Delta^{q-1} \to \Delta^q$ is defined by

$$(2.6) \qquad \varepsilon^i(t_0,\ldots,t_{q-1}) = (t_0,\ldots,t_{i-1},0,t_i,\ldots,t_{q-1}).$$

Analogously, the degeneracy operators η_i are defined by $\eta_i(\sigma) = \sigma \circ \eta^i$, $i = 0,\ldots,q$, where $\eta^i : \Delta^{q+1} \to \Delta^q$ is defined by

$$(2.7) \quad \eta^i(t_0,\ldots,t_{q+1}) = (t_0,\ldots,t_{i-1},t_i+t_{i+1},t_{i+2},\ldots,t_{q+1}).$$

We leave it to the reader to verify the above identities.

A map of simplicial sets is clearly a sequence of maps commuting with the face and degeneracy operators. Obviously S^∞ and S^{top} become functors from the category of C^∞ manifolds (respectively topological spaces) to the category of simplicial sets.

Definition 2.8. Let $S = \{S_q\}$ be a simplicial set. A differential k-form φ on S is a family $\varphi = \{\varphi_\sigma\}$, $\sigma \in \coprod_p S_p$ of k-forms such that

(i) φ_σ is a k-form on the standard simplex Δ^p for $\sigma \in S_p$.

(ii) $\varphi_{\varepsilon_i\sigma} = (\varepsilon^i)^*\varphi_\sigma$, $i = 0,\ldots,p$, $\sigma \in S_p$, $p = 1,2,\ldots$;

where $\varepsilon^i : \Delta^{p-1} \to \Delta^p$ is the i-th face map as defined by (2.6).

Example 2. Let $S = S^\infty(M)$ for M a C^∞ manifold. Then if ω is a k-form on M we get a k-form $\varphi = \{\varphi_\sigma\}$ on $S^\infty(M)$ by putting $\varphi_\sigma = \sigma^*\omega$ for $\sigma \in S_p^\infty(M)$.

The set of k-forms on a simplicial set S is denoted $A^k(S)$. If $\varphi \in A^k(S)$, $\psi \in A^l(S)$ we have again the wedge-product $\varphi \wedge \psi$ defined by

$$(2.9) \qquad (\varphi \wedge \psi)_\sigma = \varphi_\sigma \wedge \psi_\sigma, \qquad \sigma \in S_p,\ p = 0,1,\ldots$$

Also, we have the exterior differential $d : A^k(S) \to A^{k+1}(S)$ defined by

$$(2.10) \qquad (d\varphi)_\sigma = d\varphi_\sigma, \qquad \sigma \in S_p,\ \ p = 0,1,2,\ldots$$

It is obvious that $\wedge$ is again associative and graded commutative and that d satisfies

$$(2.11) \qquad dd = 0 \quad \text{and}$$

$$d(\varphi \wedge \psi) = d\varphi \wedge \psi + (-1)^k \varphi \wedge d\psi, \quad \varphi \in A^k(S),\ \psi \in A^l(S).$$

We shall call $(A^*(S),\wedge,d)$ the simplicial de Rham algebra or de Rham complex of S. If $f : S \to S'$ is a simplicial map then clearly we get $f^* : A^*(S') \to A^*(S)$ defined by

$$(2.12) \qquad (f^*\varphi)_\sigma = \varphi_{f\sigma}, \qquad \varphi \in A^k(S'),\ \sigma \in S_p,\ p = 0,1,\ldots$$

and thus A^* is a contravariant functor.

Remark 1. Notice that by Example 2 we have for any C^∞ manifold M a natural transformation

$$i : A^*(M) \to A^*(S^\infty(M)) \tag{2.13}$$

which is clearly injective, so we can think of simplicial forms on $S^\infty(M)$ as some generalized kind of forms on M.

We now want to prove a "de Rham theorem" for any simplicial set S. The chain complex $C_*(S)$ with real coefficients is of course the complex where $C_k(S)$ is the free vector space on S_k and $\partial : C_k(S) \to C_{k-1}(S)$ is given by

$$\partial(\sigma) = \sum_{i=0}^{k} (-1)^i \varepsilon_i(\sigma), \qquad \sigma \in S_k.$$

Dually the cochain complex with real coefficients is $C^*(S) = \mathrm{Hom}(C_*(S), \mathbb{R})$, so again a k-cochain is a family $c = (c_\sigma)$, $\sigma \in S_k$, and $\delta : C^k(S) \to C^{k+1}(S)$ is given by

$$(\delta c)_\sigma = \sum_{i=0}^{k+1} (-1)^i c_{\varepsilon_i \tau}, \qquad \tau \in S_{k+1}. \tag{2.14}$$

Again we have a natural map

$$I : A^k(S) \to C^k(S)$$

defined by

$$I(\varphi)_\sigma = \int_{\Delta^k} \varphi_\sigma, \qquad \varphi \in A^k(S),\ \sigma \in S_k, \tag{2.15}$$

and we can now state

Theorem 2.16 (H. Whitney). $I : A^*(S) \to C^*(S)$ is a chain map inducing an isomorphism in homology. In fact there is a natural chain map $E : C^*(S) \to A^*(S)$ and natural chain homotopies $s_k : A^k(S) \to A^{k-1}(S)$, $k = 1,2,\ldots,$ such that

$$(2.17) \qquad I \circ d = \delta \circ I, \qquad E \circ \delta = d \circ E$$

$$(2.18) \qquad I \circ E = \mathrm{id}, \qquad E \circ I - \mathrm{id} = s_{k+1} \circ d + d \circ s_k, \quad k = 0,1,\ldots$$

For the proof we first need some preparations. As usual $\Delta^p \subseteq \mathbb{R}^{p+1}$ is the standard p-simplex spanned by the canonical basis $\{e_0,\ldots,e_p\}$ and we use the barycentric coordinates $(t_0,\ldots,t_p)$. Now Δ^p is star shaped with respect to each vertex e_j, $j = 0,\ldots,p$, and therefore we have operators $h_{(j)} : A^k(\Delta^p) \to A^{k-1}(\Delta^p)$, $k = 1,2,\ldots,$ for each j as defined in the proof of Lemma 1.2. Also put $h_{(j)}\omega = 0$ for $\omega \in A^0(\Delta^p)$. The proof of the following lemma is left as an exercise (cf. Exercise 3 of Chapter 1):

<u>Lemma 2.19</u>. The operators $h_{(j)} : A^k(\Delta^p) \to A^{k-1}(\Delta^p)$, $k = 0,1,2,\ldots,$ satisfy

(i) For $\omega \in A^k(\Delta^p)$

$$(2.20) \qquad h_{(j)}d\omega + dh_{(j)}\omega = \begin{cases} -\omega, & k > 0 \\ \omega(e_j)-\omega, & k = 0 \end{cases}$$

(ii) For $i,j = 0,\ldots,p$

$$(2.21) \qquad (\varepsilon^i)^* \circ h_{(j)} = \begin{cases} h_{(j)} \circ (\varepsilon^i)^*, & i > j \\ h_{(j-1)} \circ (\varepsilon^i)^*, & i < j \end{cases}$$

(iii) For $\omega \in A^k(\Delta^k)$

$$(2.22) \qquad \int_{\Delta^k} \omega = (-1)^k h_{(k-1)} \circ \ldots \circ h_{(0)}(\omega)(e_k).$$

Next some notation: Consider a fixed integer $p \geqq 0$.

Let $I = (i_0, \ldots, i_k)$ be a sequence of integers satisfying $0 \leqq i_0 < i_1 < \ldots < i_k \leqq p$. The "dimension" of I is $|I| = k$ (for $I = \emptyset$ put $|\emptyset| = -1$). Corresponding to I we have the inclusion $\mu^I : \Delta^k \to \Delta^p$ onto the k-dimensional face spanned by $\{e_{i_0}, \ldots, e_{i_k}\}$ and similarly we have a face map $\mu_I : S_p \to S_k$. Explicitly, $\mu^I = \varepsilon^{j_1} \circ \ldots \circ \varepsilon^{j_l}$ and $\mu_I = \varepsilon_{j_l} \circ \ldots \circ \varepsilon_{j_1}$ where $p \geqq j_1 > \ldots > j_l \geqq 0$ is the complementary sequence to I and $k + l = p$. Also associated to I there is the "elementary form" ω_I on Δ^p defined by

$$(2.23) \qquad \omega_I = \sum_{s=0}^{k} (-1)^s t_{i_s} dt_{i_0} \wedge \ldots \wedge \hat{dt}_{i_s} \wedge \ldots \wedge dt_{i_k}$$

(for $I = \emptyset$ put $\omega_\emptyset = 0$) and the operator

$$h_I = h_{(i_k)} \circ \ldots \circ h_{(i_0)} : A^*(\Delta^p) \to A^*(\Delta^p)$$

which lowers the degree by $k + 1$ (for $I = \emptyset$ put $h_\emptyset = \mathrm{id}$).

We can now define $E : C^k(S) \to A^k(S)$ as follows (a motivation is given in Exercise 1 below):

For $c = (c_\tau)$ a k-cochain and $\sigma \in S_p$ put

$$(2.24) \qquad E(c)_\sigma = k! \sum_{|I|=k} \omega_I \cdot c_{\mu_I(\sigma)}$$

which is clearly a k-form on Δ^p (if $p < k$ the sum is of course interpreted as zero). Similarly $s_k : A^k(S) \to A^{k-1}(S)$ is defined as follows: For $\varphi = (\varphi_\sigma) \in A^k(S)$ and $\sigma \in S^p$ put

$$(2.25) \qquad s_k(\varphi)_\sigma = \sum_{0 \leqq |I| < k} |I|!\,\omega_I \wedge h_I(\varphi_\sigma)$$

which is clearly a k-1-form on Δ^p.

First we show that (2.24) satisfies Definition 2.8 (ii): Let $l \in \{0, \ldots, p\}$ and suppose $I = (i_0, \ldots, i_k)$ does <u>not</u> contain l. Then for some s we have $i_s < l < i_{s+1}$ and we put $I' = (i_0,\dots,i_s,i_{s+1}-1,\dots,i_k-1)$. With this notation

$$(\varepsilon^1)^* E(c)_\sigma = k! \sum_{|I|=k;\,1\notin I} \omega_{I'} \cdot c_{\mu_I(\sigma)}$$

$$= k! \sum_{|I'|=k} \omega_{I'}\, c_{\mu_{I'}(\varepsilon_1\sigma)}$$

since it is easy to see that $\mu_I(\sigma) = \mu_{I'}(\varepsilon_1\sigma)$. Now since $I' = (i'_0,\dots,i'_k)$ runs over all sequences satisfying $0 \leqq i'_0 < \dots < i'_k \leqq p-1$, the last expression above equals $E(c)_{\varepsilon_1\sigma}$ which was to be proved. Similarly (2.25) is shown to satisfy Definition 2.8 (i) using (2.21) above.

Now let us prove the identities (2.17): The first identity of (2.17) is proved exactly as Lemma 1.14, so let us concentrate on the second one: For $c \in C^k(S)$ and $\sigma \in S_p$ we have

$$(2.26)\quad dE(c)_\sigma = k! \sum_{|I|=k} \Big(\sum_{s=0}^{k} (-1)^s dt_{i_s} \wedge dt_{i_0} \wedge \dots \wedge \widehat{dt}_{i_s} \wedge \dots \wedge dt_{i_k}\Big) c_{\mu_I(\sigma)}$$

$$= (k+1)! \sum_{|I|=k} dt_{i_0} \wedge \dots \wedge dt_{i_k} \cdot c_{\mu_I(\sigma)}.$$

On the other hand

$$(2.27)\quad E(\delta c)_\sigma = (k+1)! \sum_{|I|=k+1} \omega_I \cdot (\delta c)_{\mu_I(\sigma)}$$

$$= (k+1)! \sum_{|I|=k+1} \omega_I \cdot \Big(\sum_{l=0}^{k+1} (-1)^l c_{\varepsilon_l \mu_I(\sigma)}\Big).$$

For $J = (j_0,\dots,j_k)$, $0 \leqq j_0 < \dots < j_k \leqq p$ we shall find the terms involving $c_{\mu_J(\sigma)}$ in (2.27). Now $\varepsilon_l \mu_I = \mu_J$ iff $(i_0,\dots,\hat{i}_l,\dots,i_{k+1}) = (j_0,\dots,j_k)$. The coefficient of $c_{\mu_J(\sigma)}$ in (2.27) therefore is

$$(2.28)\quad (k+1)! \sum_{i \notin (j_0,\dots,j_k)} (-1)^l \sum_{s=0}^{k+1} (-1)^s t_{i_s} dt_{i_0} \wedge \dots \wedge \widehat{dt}_{i_s} \wedge \dots \wedge dt_{i_{k+1}}$$

where $(i_0,\dots,i_{k+1}) = (j_0,\dots,j_{l-1},i,j_l,\dots,j_k)$.

Now (2.28) equals

$$(k+1)!\sum_{i\in(j_0,\dots,j_k)}\Big[\sum_{s<l}(-1)^{s+1}t_{j_s}dt_{j_0}\wedge\dots\wedge\hat{dt}_{j_s}\wedge\dots\wedge dt_{j_{l-1}}\wedge dt_i\wedge dt_{j_l}\wedge\dots$$

$$\dots\wedge dt_{j_k}+t_i dt_{j_0}\wedge\dots\wedge dt_{j_k}+$$

$$+\sum_{s\geq l}(-1)^{s+l-1}t_{j_s}dt_{j_0}\wedge\dots\wedge dt_{j_{l-1}}\wedge dt_i\wedge dt_{j_l}\wedge\dots\wedge\hat{dt}_{j_s}\wedge\dots\wedge dt_{j_k}\Big]$$

$$=(k+1)!\sum_{i\in(j_0,\dots,j_k)}\Big[t_i dt_{j_0}\wedge\dots\wedge dt_{j_k}+$$

$$+\sum_{s=0}^{k}-t_{j_s}dt_{j_0}\wedge\dots\wedge dt_{j_{s-1}}\wedge dt_i\wedge dt_{j_{s+1}}\wedge\dots\wedge dt_{j_k}\Big]$$

$$=(k+1)!\Big[\sum_{i\in(j_0,\dots,j_k)}t_i dt_{j_0}\wedge\dots\wedge dt_{j_k}+$$

$$+\sum_{s=0}^{k}\sum_{i\in(j_0,\dots,j_k)}-t_{j_s}dt_{j_0}\wedge\dots\wedge dt_{j_{s-1}}\wedge dt_i\wedge dt_{j_{s+1}}\wedge\dots\wedge dt_{j_k}\Big]$$

$$=(k+1)!\Big[\sum_{i\in(j_0,\dots,j_k)}t_i dt_{j_0}\wedge\dots\wedge dt_{j_k}+$$

$$+\sum_{s=0}^{k}t_{j_s}dt_{j_0}\wedge\dots\wedge dt_{j_k}-\sum_{s=0}^{k}t_{j_s}dt_{j_0}\wedge\dots\wedge dt_{j_{s-1}}\wedge\Big(\sum_{i=0}^{p}dt_i\Big)\wedge dt_{j_{s+1}}\wedge\dots\wedge dt_{j_k}\Big]$$

$$=(k+1)!\sum_{i=0}^{p}t_i dt_{j_0}\wedge\dots\wedge dt_{j_k}=(k+1)!dt_{j_0}\wedge\dots\wedge dt_{j_k}$$

since $\sum_{i=0}^{p}dt_i=0$ and $\sum_{i=0}^{p}t_i=1$. Hence

$$E(\delta c)_\sigma=(k+1)!\sum_{|J|=k}dt_{j_0}\wedge\dots\wedge dt_{j_k}\cdot c_{\mu_J(\sigma)}=dE(c)_\sigma$$

by (2.26) which proves the second identity of (2.17).

To prove the first equation of (2.18) consider a k-cochain $c=(c_\sigma)$, $\sigma\in S_k$ and we shall show that $I(E(c))_\sigma=c_\sigma$. By (2.24) $E(c)_\sigma$ is the k-form on Δ^k given by

$$E(c)_\sigma = k!c_\sigma \sum_{j=0}^{k} (-1)^j t_j dt_0\wedge\ldots\wedge d\hat{t}_j\wedge\ldots\wedge dt_k$$

$$= k!c_\sigma[t_0 dt_1\wedge\ldots\wedge dt_k +$$

$$+ \sum_{j=1}^{k} (-1)^j t_j (-\sum_{s=1}^{k} dt_s)\wedge dt_1\wedge\ldots\wedge d\hat{t}_j\wedge\ldots\wedge dt_k]$$

$$= k!c_\sigma[t_0 dt_1\wedge\ldots\wedge dt_k + \sum_{j=1}^{k} (-1)^{j-1} t_j dt_j\wedge dt_1\wedge\ldots\wedge d\hat{t}_j\wedge\ldots\wedge dt_k]$$

$$= k!c_\sigma dt_1\wedge\ldots\wedge dt_k.$$

Therefore

$$I(E(c))_\sigma = k!c_\sigma \int_{\Delta^k} dt_1\wedge\ldots\wedge dt_k = c_\sigma$$

by Exercise 1 of Chapter 1.

For the proof of the second equation of (2.18) first observe that an iterated application of (2.20) yields the following

Lemma 2.29. Let $\omega \in A^k(\Delta^p)$, $k \geq 0$, and consider $I = (i_0,\ldots,i_r)$, $0 \leq r \leq p$, with $0 \leq i_0 <\ldots< i_r \leq p$. Suppose $k \geq r$. Then

$$h_I(d\omega) = \begin{cases} -\sum_{j=0}^{r} (-1)^j h_{(i_0,\ldots,\hat{i}_j,\ldots,i_r)}(\omega) - (-1)^r dh_I(\omega), & k > r \\ -\sum_{j=0}^{k} (-1)^j h_{(i_0,\ldots,\hat{i}_j,\ldots,i_k)}(\omega) + (-1)^k h_{(i_0,\ldots,i_{k-1})}(\omega)(e_{i_k}), & k = r. \end{cases}$$

Now let $\varphi \in A^k(S)$ and $\sigma \in S_p$. Assume $p \geq k$ (otherwise there is nothing to prove). By (2.29)

$$(2.30)\quad s_{k+1}(d\varphi)_\sigma = \sum_{0\leq|I|\leq k} |I|!\omega_I \wedge h_I(d\varphi_\sigma)$$

$$= \sum_{|I|=k} k!\omega_I \cdot ((-1)^k h_{(i_0,\dots,i_{k-1})}(\varphi_\sigma)(e_{i_k}))$$

$$- \sum_{0\leq|I|\leq k} |I|!\omega_I \wedge (\sum_{j=0}^{|I|}(-1)^j h_{(i_0,\dots,\hat{i}_j,\dots,i_{|I|})}(\varphi_\sigma))$$

$$- \sum_{0\leq|I|<k} |I|!\omega_I \wedge ((-1)^{|I|} dh_I(\varphi_\sigma)).$$

Also

$$(2.31)\quad d(s_k\varphi)_\sigma = \sum_{0\leq|I|<k} |I|!d\omega_I \wedge h_I(\varphi_\sigma)+(-1)^{|I|}|I|!\omega_I\wedge dh_I(\varphi_\sigma).$$

By (2.22)

$$(-1)^k h_{(i_0,\dots,i_{k-1})}(\varphi_\sigma)(e_{i_k}) = \int_{\Delta^k}(\mu^I)^*\varphi_\sigma = \int_{\Delta^k}\varphi_{\mu_I(\sigma)}$$

$$= I(\varphi)_{\mu_I(\sigma)}.$$

Therefore adding (2.30) and (2.31) we obtain

$$(2.32)\quad s_{k+1}(d\varphi)_\sigma + d(s_k\varphi)_\sigma =$$

$$= E(I(\varphi))_\sigma - \varphi_\sigma - \sum_{0<|I|\leq k} |I|!\omega_I\wedge(\sum_{j=0}^{|I|}(-1)^j h_{(i_0,\dots,\hat{i}_j,\dots,i_{|I|})}(\varphi_\sigma))$$

$$+ \sum_{0\leq|I|<k} |I|!d\omega_I \wedge h_I(\varphi_\sigma).$$

However the last two sums in (2.32) cancel by exactly the same calculations as in the proof that (2.26) equals (2.27) above. This proves the second equation of (2.18) and ends the proof of Theorem 2.16.

We now return to the proof of Theorem 2.4. Notice that in the commutative diagram

$$\begin{array}{ccc} A^*(M) & \xrightarrow{\quad i \quad} & A^*(S^\infty(M)) \\ & {\scriptstyle I}\searrow \quad \swarrow{\scriptstyle I} & \\ & C^*(M) & \end{array}$$

all maps induce isomorphism in homology. Also $i : A^*(M) \to A^*(S^\infty(M))$ is obviously multiplicative. Theorem 2.4 therefore immidiately follows from

Theorem 2.33. For any simplicial set S the following diagram commutes

$$\begin{array}{ccc} H(A^*(S)) \otimes H(A^*(S)) & \xrightarrow{\quad \wedge \quad} & H(A^*(S)) \\ \downarrow I \otimes I & & \downarrow I \\ H(C^*(S)) \otimes H(C^*(S)) & \xrightarrow{\quad \smile \quad} & H(C^*(S)) \end{array}$$

where the upper horizontal map is induced by the wedge-product of simplicial forms and the lower horizontal map is the cup-product.

Before proving this theorem let us recall the definition of the cup-product in $H(C^*(S))$.

Consider the functor C_* from the category of simplicial sets to the category of chain-complexes and chain maps (as usual we take coefficients equal to $\mathbb{R}$). An approximation to the diagonal is a natural transformation

$$\Phi : C_*(S) \to C_*(S) \otimes C_*(S)$$

(in particular a chain map) such that in dimension zero Φ is given by

$$\Phi(\sigma) = \sigma \otimes \sigma, \quad \sigma \in S_0.$$

It follows using acyclic models that there exists some Φ and it is unique up to chain homotopy (see e.g. A. Dold [10, Chapter 6, § 11, Exercise 4]. The cup-product is now simply induced by the composed mapping

$$\Phi^* : C^*(S) \otimes C^*(S) \to \operatorname{Hom}(C_*(S) \otimes C_*(S), \mathbb{R}) \to C^*(S).$$

An explicit choice for Φ is the Alexander-Whitney map AW defined by

$$(2.34) \qquad AW(\sigma) = \sum_{p=0}^{n} \mu_{(0,\dots,p)}(\sigma) \otimes \mu_{(p,\dots,n)}(\sigma), \quad \sigma \in S_n.$$

With this choice of Φ the cup-product is explicitly given as follows: Let $a = (a_\sigma) \in C^p(S)$ and $b = (b_\tau) \in C^q(S)$; then $a \smile b$ is represented by the cochain

$$(2.35) \qquad (a \smile b)_\sigma = a_{\mu(0,\dots,p)(\sigma)} \cdot b_{\mu(p,\dots,p+q)(\sigma)}, \quad \sigma \in S_{p+q}.$$

<u>Proof of Theorem 2.32</u>. By Theorem 2.16 every simplicial form is cohomologous to a form in the image of $\mathcal{E} : C^*(S) \to A^*(S)$. It is therefore enough to show that for $a \in C^p(S)$, $b \in C^q(S)$ the (p+q)-cochain $\mathcal{I}(\mathcal{E}(a) \wedge \mathcal{E}(b))$ represents the cup-product of a and b in $H(C^*(S))$. So let $\Phi^* : C^*(S) \otimes C^*(S) \to C^*(S)$ be defined by

$$(2.36) \qquad \Phi^*(a \otimes b) = \mathcal{I}(\mathcal{E}(a) \wedge \mathcal{E}(b)), \quad a \in C^p(S),\ b \in C^q(S).$$

We claim that there is an approximation to the diagonal Φ inducing (2.36). Let us find an explicit formula for (2.36):

Put $n = p + q$ and consider $\sigma \in S_n$. Then on Δ^n,

$$\mathcal{E}(a)_\sigma = p! \sum_{|I|=p} a_{\mu_I(\sigma)} \omega_I, \qquad \mathcal{E}(b)_\sigma = q! \sum_{|J|=q} b_{\mu_J(\sigma)} \omega_I,$$

where as usual $I = (i_0,\dots,i_p)$ and $J = (j_0,\dots,j_q)$ satisfy $0 \leqq i_0 <\dots< i_p \leqq n$, $0 \leqq j_0 <\dots< j_q \leqq n$. Then I and J has at least one integer in common. If I and J has more than two integers in common then obviously $\omega_I \wedge \omega_J = 0$. Now suppose I and J have exactly <u>two</u> integers in common, say

$i_{s_1} = j_{r_1} < i_{s_2} = j_{r_2}$. Then

$$\omega_I \wedge \omega_J = (-1)^{s_1+r_1} t_{i_{s_1}} t_{j_{r_2}} dt_{i_0} \wedge \ldots \wedge \hat{dt}_{i_{s_1}} \wedge \ldots \wedge dt_{i_p} \wedge dt_{j_0} \wedge \ldots \wedge \hat{dt}_{j_{r_2}} \wedge \ldots \wedge dt_{j_q}$$

$$+ (-1)^{s_2+r_2} t_{i_{s_2}} t_{j_{r_1}} dt_{i_0} \wedge \ldots \wedge \hat{dt}_{i_{s_2}} \wedge \ldots \wedge dt_{i_p} \wedge dt_{j_0} \wedge \ldots \wedge \hat{dt}_{j_{r_1}} \wedge \ldots \wedge dt_{j_q}$$

and it is easy to see that these two terms are equal with opposite signs so $\omega_I \wedge \omega_J = 0$ also in this case. Finally suppose I and J have exactly one integer in common, say $i_s = j_r$; then

$$\omega_I \wedge \omega_J = (-1)^{s+r} t_{i_s} t_{j_r} dt_{i_0} \wedge \ldots \wedge \hat{dt}_{i_s} \wedge \ldots \wedge dt_{i_p} \wedge dt_{j_0} \wedge \ldots \wedge \hat{dt}_{j_r} \wedge \ldots \wedge dt_{j_q}$$

$$+ \sum_{k \neq r} (-1)^{s+k} t_{i_s} t_{j_k} dt_{i_0} \wedge \ldots \wedge \hat{dt}_{i_s} \wedge \ldots \wedge dt_{i_p} \wedge dt_{j_0} \wedge \ldots \wedge \hat{dt}_{j_k} \wedge \ldots \wedge dt_{j_q}$$

$$+ \sum_{l \neq s} (-1)^{r+l} t_{i_l} t_{j_r} dt_{i_0} \wedge \ldots \wedge \hat{dt}_{i_l} \wedge \ldots \wedge dt_{i_p} \wedge dt_{j_0} \wedge \ldots \wedge \hat{dt}_{j_r} \wedge \ldots \wedge dt_{j_q} .$$

Using $\sum_{\lambda=0}^{n} dt_\lambda = 0$ we get

$$\omega_I \wedge \omega_J = [(-1)^{s+r} t_{i_s} t_{j_r} + \sum_{k \neq r} (-1)^{s+k+r+k} t_{i_s} t_{j_k} + \sum_{l \neq s} (-1)^{r+l+l+s} t_{i_l} t_{j_r}] \cdot$$

$$\cdot\, dt_{i_0} \wedge \ldots \wedge \hat{dt}_{i_s} \wedge \ldots \wedge dt_{i_p} \wedge dt_{j_0} \wedge \ldots \wedge \hat{dt}_{j_r} \wedge \ldots \wedge dt_{j_q}$$

$$= (-1)^{s+r} t_{i_s} dt_{i_0} \wedge \ldots \wedge \hat{dt}_{i_s} \wedge \ldots \wedge dt_{i_p} \wedge dt_{j_0} \wedge \ldots \wedge \hat{dt}_{j_r} \wedge \ldots \wedge dt_{j_q} .$$

It follows that

$$(E(a) \wedge E(b))_\sigma = p!q! \sum_{\substack{|I|=p \\ |J|=q}} a_{\mu_I(\sigma)} b_{\mu_J(\sigma)} \cdot (-1)^{r+s} t_{i_s} dt_{i_0} \wedge \ldots \wedge \hat{dt}_{i_s} \wedge \ldots$$

$$\ldots \wedge dt_{i_p} \wedge dt_{j_0} \wedge \ldots \wedge \hat{dt}_{j_r} \wedge \ldots \wedge dt_{j_q}$$

where the sum is taken over I and J such that for some s

and r $i_s = j_r$ and no other integers are common. Now let $\text{sgn}(I,J)$ be the sign $(-1)^{p-s+r}$ · times the sign of the permutation taking $(0,\dots,n)$ into $(i_0,\dots,\hat{i}_s,\dots,i_p,i_s=j_r,j_0,\dots,\hat{j}_r,\dots,j_q)$; then

$$\text{sgn}(I,J)\int_{\Delta^n}(-1)^{r+s}t_{i_s}dt_{i_0}\wedge\dots\wedge d\hat{t}_{i_s}\wedge\dots\wedge dt_{i_p}\wedge dt_{j_0}\wedge\dots\wedge d\hat{t}_{j_r}\wedge\dots\wedge dt_{j_q}$$

$$= \int_{\Delta^n}t_0dt_1\wedge\dots\wedge dt_n = \int_{\{t_1+\dots+t_n\leqq 1,\, t_i\geqq 0\}}(1-(t_1+\dots+t_n))dt_1dt_2\dots dt_n$$

$$= \int_{\{t_0+\dots+t_n\leqq 1,\, t_i\geqq 0\}}dt_0\dots dt_n = \int_{\Delta^{n+1}}dt_1\wedge dt_2\wedge\dots\wedge dt_{n+1} = 1/(n+1)!\ .$$

Hence

$$(2.37)\qquad \Phi^*(a\otimes b)_\sigma = I(E(a)\wedge E(b))_\sigma$$

$$= \frac{p!q!}{(p+q+1)!}\sum_{\substack{|I|=p\\|J|=q}}\text{sgn}(I,J)\,a_{\mu_I(\sigma)}b_{\mu_J(\sigma)}$$

where again I and J have exactly one integer in common. Therefore if we define the map

$$\Phi : C_*(S)\to C_*(S)\otimes C_*(S)$$

by

$$(2.38)\qquad \Phi(\sigma) = \sum_{p+q=n}\frac{p!q!}{(n+1)!}\sum_{\substack{|I|=p\\|J|=q}}\text{sgn}(I,J)\mu_I(\sigma)\otimes\mu_J(\sigma),\ \sigma\in S_n$$

then Φ^* given by (2.36) is the dual map. We want to show that Φ is an approximation to the diagonal: Clearly Φ is natural and

$$\Phi(\sigma) = \sigma\otimes\sigma \quad\text{for}\quad \sigma\in S_0.$$

It remains to show that Φ is a chain map. However, for this it is enough to see that Φ^* is a chain map which is easy:

$$\begin{aligned}\Phi^*(\delta(a\otimes b)) &= \Phi^*(\delta a\otimes b + (-1)^p\, a\otimes \delta b)\\ &= I(E(\delta a)\wedge E(b)) + (-1)^p I(E(a)\wedge E(\delta b))\\ &= I(dE(a)\wedge E(b) + (-1)^p E(a)\wedge dE(\delta b))\\ &= I(d(E(a)\wedge E(b)) = \delta I(E(a)\wedge E(b)) = \delta\Phi^*(a\otimes b).\end{aligned}$$

This ends the proof.

Remark. Notice that the term in (2.37) corresponding to $I = (0,\ldots,p)$, $J = (p,\ldots,p+q)$ gives exactly the Alexander-Whitney cup-product (2.35). Thus (2.37) is an average of the Alexander-Whitney cup-product over the permutations given by (I,J) in order to make the product graded commutative on the cochain level. In fact the A-W-product is not graded commutative on the cochain level as Φ^* clearly must be since $\wedge$ is graded commutative. On the other hand the A-W-product is associative on the cochain level which Φ^* is not. In order to achieve both properties it seems necessary to replace the functor C^* by the chain equivalent functor A^*.

Exercise 1. Consider for $k < p$ a sequence $I = (i_0,\ldots,i_k)$ with $0 \leqq i_0 <\ldots< i_k \leqq p$ and let $\Delta_I^p \subseteqq \Delta^p \subseteqq \mathbb{R}^{p+1}$ be the set

$$\Delta_I^p = \{(t_0,\ldots,t_p) \mid \text{some } t_{i_s} > 0\} = \Delta^p - \{t_{i_0} = t_{i_1} = \ldots = t_{i_k} = 0\},$$

(i.e. we subtract a $p-k-1$-dimensional face). Let $\pi_I : \Delta_I^p \to \Delta^k$ be the projection

$$\pi_I(t_0,\ldots,t_p) = \frac{1}{\sum_s t_{i_s}}\,(t_{i_0},\ldots,t_{i_k}).$$

a) Show that on Δ_I^p

$$\pi_I^*(dt_1\wedge\ldots\wedge dt_k) = \Big(\sum_s t_{i_s}\Big)^{-(k+1)}\omega_I$$

where ω_I is given by (2.23).

b) Show the following properties of ω_I:

(i) $(\mu^I)^*\omega_I = dt_1 \wedge \ldots \wedge dt_k$

(ii) $(\mu^J)^*\omega_I = 0$ if $|J| = k$, $J \neq I$.

c) Conclude that for $c = (c_\sigma)$ a k-cochain and $\sigma \in S_p$, the form $E(c)_\sigma$ on Δ^p satisfy: For any $I = (i_0,\ldots,i_k)$ as above

(2.39) $$(\mu^I)^*E(c)_\sigma = k!c_{\mu_I(\sigma)}dt_1 \wedge \ldots \wedge dt_k .$$

d) Observe that for $\sigma \in S_k$ the k-form on Δ^k

$$E(c)_\sigma = k!c_\sigma dt_1 \wedge \ldots \wedge dt_k$$

is the simplest choice in order to satisfy the first identity of (2.18). Show that with this choice for $\sigma \in S_k$ the condition (2.39) is a necessary requirement for the choice of $E(c)_\sigma$ for $\sigma \in S_p$, $p > k$.

<u>Exercise 2</u>. a) Let $f : S \to S'$ be a simplicial map of simplicial sets. Show that

(i) $I \circ f^* = f^{\#} \circ I$

(ii) $f^* \circ E = E \circ f^{\#}$

(iii) $s_k \circ f^* = f^* \circ s_k$, $k = 1,2,\ldots$

b) Two simplicial maps $f_0, f_1 : S \to S'$ are called <u>homotopic</u> if for each q there are functions $h_i : S_q \to S'_{q+1}$, $i = 0,\ldots,q$, such that

(i) $\varepsilon_0 h_0 = f_0$, $\varepsilon_{q+1}h_q = f_1$

(ii) $$\varepsilon_i h_j = \begin{cases} h_{j-1}\varepsilon_i, & \text{if } i < j, \\ h_j \varepsilon_{i-1}, & \text{if } i > j+1, \end{cases}$$

$$\varepsilon_{j+1}h_{j+1} = \varepsilon_{j+1}h_j ,$$

$$\text{(iii)}\quad \eta_i h_j = \begin{cases} h_{j+1}\eta_i, & \text{if } i \leq j, \\ h_j \eta_{i-1}, & \text{if } i > j. \end{cases}$$

Show that $f_0^{\#}, f_1^{\#} : C^*(S') \to C^*(S)$ are chain homotopic.

c) Let $f_0, f_1 : S \to S'$ be homotopic. Show that a) and b) imply that $f_0^*, f_1^* : A^*(S') \to A^*(S)$ are chain homotopic.

d) Find explicit chain homotopies in c).

<u>Exercise 3</u>. Let S be a simplicial set. A k-form $\varphi = \{\varphi_\sigma\}$ on S is called <u>normal</u> if it furthermore satisfies

$$\text{(iii)}\quad \varphi_{\eta_i\sigma} = (\eta^i)^*\varphi_\sigma, \quad i = 0,\ldots,p,\ \sigma \in S_p,\ p = 0,1,2,\ldots$$

where $\eta^i : \Delta^{p+1} \to \Delta^p$ is the i-th degeneracy map defined by (2.7). Let $A_N^k(S) \subseteq A^k(S)$ be the subset of normal k-forms on S.

a) Show that d and $\wedge$ preserve normal forms and if $f : S \to S'$ is a simplicial map then f^* also preserves normal forms.

b) Show that the operators $h_{(j)} : A^k(\Delta^p) \to A^{k-1}(\Delta^p)$, $k = 0,1,\ldots,\ j = 0,\ldots,p,$ satisfy

$$\text{(i)}\quad h_{(i)}\eta_j^* = \begin{cases} \eta_j^* h_{(i)}, & i \leq j \\ \eta_j^* h_{(i-1)}, & i > j \end{cases}$$

$$\text{(ii)}\quad h_{(i)}h_{(i)} = 0, \quad i = 0,\ldots,p.$$

c) Let $C_N^k(S) \subseteq C^k(S)$ be the set of <u>normal</u> cochains, i.e., k-cochains $c = (c_\sigma)$ such that $c_{\eta_i\tau} = 0 \ \ \forall \tau \in S_{k-1}$, $i = 0,\ldots,k-1$. Show that

(i) $I : A_N^*(S) \to C_N^*(S)$

(ii) $E : C_N^*(S) \to A_N^*(S)$

(iii) $s_k : A_N^k(S) \to A_N^{k-1}(S)$

and conclude that $I : A_N^*(S) \to C_N^*(S)$ is a chain equivalence. Hence since the inclusion $C_N^*(S) \to C^*(S)$ is a chain equivalence (see e.g. S. MacLane [18, Chapter 7, § 6] also the inclusion $A_N^*(S) \to A^*(S)$ is a chain equivalence.

Exercise 4. (D. Sullivan). Let $A^k(\Delta^n, \mathbb{Q})$ denote the set of polynomial forms with rational coefficients, i.e. $\omega \in A^k(\Delta^n, \mathbb{Q})$ is the restriction of a k-form in $\mathbb{R}^{n+1}$ of the form

$$\omega = \sum_{i_0<\ldots<i_k} a_{i_0\ldots i_k} dt_{i_0} \wedge \ldots \wedge dt_{i_k}$$

where $a_{i_0\ldots i_k}$ are polynomials in $t_0 \ldots t_n$ with rational coefficients.

Now let S be a simplicial set. A k-form $\varphi = \{\varphi_\sigma\}$ on S is called rational if $\varphi_\sigma \in A^k(\Delta^p, \mathbb{Q})$ for $\sigma \in S_p$. Let $A^k(S, \mathbb{Q})$ denote the set of rational k-forms.

a) Show that $A^*(S, \mathbb{Q})$ is a rational vector space which is closed under the exterior differential d and exterior multiplication $\wedge$.

b) Let $C^*(S, \mathbb{Q})$ denote the complex of cochains with rational values. Show that

(i) $I : A^*(S, \mathbb{Q}) \to C^*(S, \mathbb{Q})$

(ii) $E : C^*(S, \mathbb{Q}) \to A^*(S, \mathbb{Q})$

(iii) $s_k : A^k(S, \mathbb{Q}) \to A^{k-1}(S, \mathbb{Q})$

and conclude that the Theorems 2.16 and 2.33 hold with $A^*(S)$ and $C^*(S)$ replaced by $A^*(S, \mathbb{Q})$ and $C^*(S, \mathbb{Q})$.

c) Formulate and prove a normal version of question b) (see Exercise 3).

Note. For a simplicial complex the construction of the simplicial de Rham complex goes back to H. Whitney [35, Chapter 7].

3. Connections in principal bundles

The theory of connections originates from the concept of "parallel translation" in a Riemannian manifold. So for motivation consider the tangent bundle TM of a differentiable manifold M; or more generally a real vector bundle V over M of dimension n. Given points $p,q \in M$ and a vector $v \in V_p$ one wants a concept of the corresponding "parallel" vector $\tau(v) \in V_q$, i.e. we require an isomorphism $\tau : V_p \to V_q$. However, unless V is a trivial bundle this seems to be an impossible requirement. What is possible is something weaker: the concept of parallel translation along a curve from p to q, that is, suppose $\gamma : [a,b] \to M$ is a differentiable curve from $\gamma(a) = p$ to $\gamma(b) = q$ and let $v \in V_p$ be a given vector; then a "connection" will associate to these data a differentiable family $v_t \in V_{\gamma(t)}$, $t \in [a,b]$, with $v_a = v$. It is of course enough to parallel translate a basis or frame $\{v_1,\ldots,v_n\}$ for the vector space V_p. Therefore let $\pi : F(V) \to M$ denote the frame bundle over M, i.e. the bundle whose fibre over p is equal to the set of all bases (frames) for V_p. Then a "connection" simply associates to any curve $\gamma : [a,b] \to M$ and any point $e \in F(V)_{\gamma(a)}$ a lift of γ through e, that is, a curve $\bar{\gamma} : [a,b] \to F(V)$ with $\bar{\gamma}(a) = e$ and $\pi \circ \bar{\gamma} = \gamma$. Now let q tend to p; then γ defines a tangent vector $X \in T_p(M)$ and $\bar{\gamma}$ defines a tangent vector $\bar{X} \in T_e(F(V))$ such that $\pi_* \bar{X} = X$. So infinitessimally a "connection" defines a "horizontal" subspace $H_e \subseteq T_e(F(V))$ mapping isomorphically onto $T_{\pi(e)}(M)$ for every $e \in F(V)$. And that is actually how we are going to define a connection formally below. Notice that $F(V)$ is the principal $Gl(n,\mathbb{R})$-bundle associated to V. So first let us recall the fundamental facts about principal G-bundles for any Lie group G. Let M be a C^∞ manifold.

Definition 3.1. A *principal G-bundle* is a differentiable mapping $\pi : E \to M$ of differentiable manifolds together with a differentiable right G-action $E \times G \to E$ satisfying

(i) For every $p \in M$ $E_p = \pi^{-1}(p)$ is an orbit.

(ii) (Local triviality) Every point of M has an open neighbourhood U and a diffeomorphism $\varphi : \pi^{-1}(U) \to U \times G$, such that

(a) the diagram

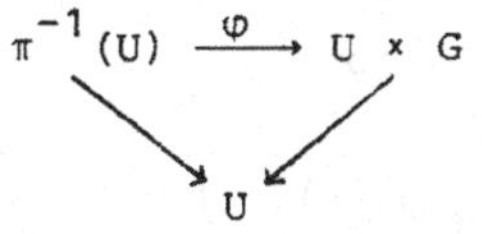

commutes,

(b) φ is equivariant, i.e.

$$\varphi(e \cdot g) = \varphi(e) \cdot g, \quad e \in \pi^{-1}(U), \ g \in G,$$

where G acts trivially on U and by right translation on G.

E is called the *total space*, M the *base space* and $E_p = \pi^{-1}(p)$ is the *fibre* at p. Notice that by (i) π is onto and by (ii) it is an open mapping so π induces a homeomorphism of the orbit space E/G to M. Also observe that the action of G on E is free (i.e., $xg = x \Rightarrow g = 1$) and the mapping $G \to E_p$ given by $g \mapsto eg$ is a diffeomorphism for every $e \in E_p$. We shall often refer to a principal G-bundle by just writing its total space E.

Example 1. Suppose $V \to M$ is an n-dimensional vector bundle. Then the bundle $F(V) \to M$ of n-frames is a principal $Gl(n,\mathbb{R})$-bundle.

Let $E \to M$ and $F \to M$ be two principal G-bundles. Then an isomorphism $\varphi : E \to F$ is a G-equivariant fibre preserving diffeomorphism. $M \times G$ is of course a trivial principal G-bundle and an isomorphism $\varphi : E \to M \times G$ is called a trivialization. The mapping φ in (ii) above is called a local trivialization.

Now consider a principal G-bundle $\pi : E \to M$ and choose a covering $\mathcal{U} = \{U_\alpha\}_{\alpha \in \Sigma}$ of M together with trivializations $\varphi_\alpha : \pi^{-1}(U_\alpha) \to U_\alpha \times G$. Then if $U_\alpha \cap U_\beta \neq \emptyset$ consider

$$\varphi_\beta \circ \varphi_\alpha^{-1} : U_\alpha \cap U_\beta \times G \to U_\alpha \cap U_\beta \times G$$

which is easily seen to be of the form

$$\varphi_\beta \circ \varphi_\alpha^{-1}(p,a) = (p, g_{\beta\alpha}(p)\cdot a), \quad a \in G,\ p \in U_\alpha \cap U_\beta$$

where $g_{\beta\alpha} : U_\alpha \cap U_\beta \to G$ is a C^∞ function. This system $\{g_{\beta\alpha}\}$ are called the transition functions for E with respect to $\mathcal{U}$ and they clearly satisfy the cocycle condition

$$(3.2) \qquad g_{\gamma\beta}(p) \cdot g_{\beta\alpha}(p) = g_{\gamma\alpha}(p), \quad \forall p \in U_\alpha \cap U_\beta \cap U_\gamma$$
$$g_{\alpha\alpha} = 1.$$

On the other hand given a covering $\mathcal{U} = \{U_\alpha\}$ and a system of transition functions satisfying (3.2) one can construct a corresponding principal G-bundle as follows: the total space is the quotient space of $\coprod_\alpha U_\alpha \times G$ with the identifications

$$(p,a) \in U_\alpha \times G \text{ identified with } (p, g_{\beta\alpha}(p)\cdot a) \in U_\beta \times G$$
$$\forall p \in U_\alpha \cap U_\beta,\ a \in G.$$

Again let $\pi : E \to M$ be a principal G-bundle and let $f : N \to M$ be a differentiable map. The "pull-back" $f^*\pi : f^*E \to N$ is the principal G-bundle with total space $f^*E \subseteqq N \times E$

$$f^*E = \{(q,e) \mid f(q) = \pi(e)\}$$

and projection $f^*\pi$ given by the restriction of the projection onto the first factor. The projection onto the second factor gives an equivariant map $\bar{f} : f^*E \to E$ covering f, i.e. the diagram

$$\begin{array}{ccc} f^*(E) & \xrightarrow{\bar{f}} & E \\ {\scriptstyle f^*\pi}\downarrow & & \downarrow{\scriptstyle \pi} \\ N & \xrightarrow{\bar{f}} & M \end{array}$$

commutes.

<u>Exercise 1</u>. a) Show that if $\{g_{\alpha\beta}\}$ is the set of transition functions for E relative to the covering $\mathcal{U} = \{U_\alpha\}_{\alpha\in\Sigma}$ then $\{g_{\alpha\beta} \circ f\}$ is the set of transition functions for f^*E relative to the covering $f^{-1}\mathcal{U} = \{f^{-1}U_\alpha\}_{\alpha\in\Sigma}$.

b) Let $F \to N$, $E \to M$ be principal G-bundles. A <u>bundle map</u> is a pair $(\bar{f},f)$, where $f : N \to M$ is a differentiable map and $\bar{f} : F \to E$ is an equivariant differentiable map covering f. Show that any bundle map factorizes into an isomorphism $\varphi : F \to f^*E$ and the canonical bundle map $f^*(E) \to E$ as above.

<u>Exercise 2</u>. a) Show that a principal G-bundle $\pi : E \to M$ is trivial iff it has a section, i.e. a differentiable map $s : M \to E$ such that $\pi \circ s = \mathrm{id}$.

b) Let $\pi : E \to M$ be a principal G-bundle. Show that π^*E is trivial.

c) Let $\pi : E \to M$ be a principal G-bundle and let $H \subseteqq G$

be a <u>closed</u> subgroup. Show that $E \to E/H$ is a principal H-bundle. (Hint: First construct local sections of the bundle $G \to G/H$ using the exponential map).

<u>Exercise 3</u>. Let $\pi : E \to M$ be a principal G-bundle and let N be a manifold with a <u>left</u> G-action $G \times N \to N$. The <u>associated fibre bundle with fibre</u> N is the mapping $\pi_N : E_N \to M$ where $E_N = E \times_G N$ is the orbit space of $E \times N$ under the G-action $(e,x)\cdot g = (eg, g^{-1}x)$, $e \in E$, $x \in N$, $g \in G$, and where π_N is induced by the projection on E followed by π. Show that E_N is a manifold and that the fibre bundle is <u>locally trivial</u> in the sense that every point of M has a neighbourhood U with a diffeomorphism $\varphi : \pi_N^{-1}(U) \to U \times N$ such that the diagram

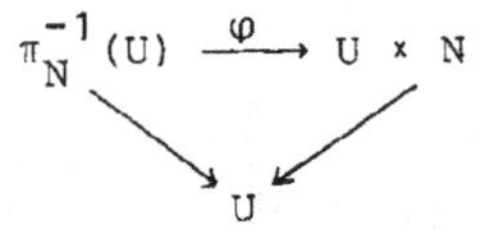

commutes. In particular π_N is open and differentiable.

Now let H and G be two Lie-groups and let $\alpha : H \to G$ be a homomorphism of Lie groups. Suppose $\pi : F \to M$ is a principal H-bundle and $\xi : E \to M$ is a principal G-bundle and suppose there is a differentiable map $\varphi : F \to E$ satisfying

$$\varphi(F_p) \subseteq E_p, \quad \forall p \in M, \text{ and}$$
$$\varphi(x \cdot h) = \varphi(x) \cdot \alpha(h), \quad \forall x \in F,\ h \in H.$$

Then we will say that E is an <u>extension</u> of F to G <u>relative</u> to α or, equivalently, that F is a <u>reduction</u> of E to H relative to α (when it is clear what α is we will omit "relative to α").

<u>Example 2</u>. An n-dimensional vector bundle $V \to M$ has the principal $Gl(N,\mathbb{R})$-bundle $F(V) \to M$. Notice that $Gl(n,\mathbb{R})$ act on the left on $\mathbb{R}^n$ and that the associated fibre bundle with fibre $\mathbb{R}^n$ is just the vector bundle. Hence there is a one-to-one correspondance between principal $Gl(n,\mathbb{R})$-bundles and vector bundles. A Riemannian metric on V defines a reduction of $F(V) \to M$ to the orthogonal group $O(n)$. In fact let $F_O(V) \subseteq F(V)$ consist of the <u>orthonormal frames</u> in each fibre. Then $F_O(V) \to M$ is the corresponding orthogonal bundle and the inclusion $F_O(V) \subseteq F(V)$ defines the reduction. Conversely a reduction of $F(V)$ to $O(n)$ clearly gives rise to a Riemannian metric on V.

<u>Exercise 4</u>. a) Let $\pi : F \to M$ be a principal H-bundle and consider G with the left H-action given by $h \cdot g = \alpha(h)g$, $h \in H$, $g \in G$. Show that the associated fibre bundle with fibre G, $\pi_G : F_G \to M$ is a G-extension of $\pi : F \to M$, and show that an extension is unique.

b) Show that a principal G-bundle $\pi : E \to M$ has a reduction to H relative to α iff there is a covering $\mathcal{U} = \{U_\gamma\}$ and a set of transition functions for E of the form $\{\alpha \circ h_{\beta\gamma}\}$ with $\{h_{\beta\gamma}\}$ a set of functions satisfying

$$(3.2) \qquad (h_{\beta\gamma} : U_\gamma \cap U_\beta \to H).$$

Before we introduce the notion of a connection in a principal bundle it is convenient to consider differential forms with coefficients in a vector space. So let M be a C^∞ manifold and V a finite dimensional vectorspace. A differential form ω on M of degree k <u>with values in V</u> associates a C^∞ function $\omega(X_1,\ldots,X_k) : M \to V$ to every set of C^∞ vector fields $X_1,\ldots,X_k$ on M; ω is again multilinear and alternating and has the "tensor property" as before. If we choose a basis $\{e_1,\ldots,e_n\}$ for V then ω is of the form $\omega = \omega_1 e_1 + \ldots + \omega_n e_n$ where $(\omega_1,\ldots,\omega_n)$ is a set of usual k-forms. Let $A^k(M,V)$ denote the set of k-forms on M with values in V. Again $A^*(M,V)$ has an exterior differential d defined by the same formula as in Chapter 1 and $A^*(M,V)$ is a chain complex (that is, $dd = 0$). This time, however, the wedge-product is a map

$$A^k(M,V) \otimes A^l(M,W) \to A^{k+l}(M,V \otimes W)$$

for V,W two vectorspaces. In fact for $\omega_1 \in A^k(M,V)$ and $\omega_2 \in A^l(M,W)$ define $\omega_1 \wedge \omega_2 \in A^{k+l}(M,V \otimes W)$ by

$$\omega_1 \wedge \omega_2(X_1,\ldots,X_{k+l})$$
$$= \frac{1}{(k+l)!} \sum_\sigma \operatorname{sign}(\sigma)\omega_1(X_{\sigma(1)},\ldots,X_{\sigma(k)}) \otimes \omega_2(X_{\sigma(k+1)},\ldots,X_{\sigma(k+l)})$$

where as usual σ runs through all permutations of $1,\ldots,k+l$. Again we have the formula

$$d(\omega_1 \wedge \omega_2) = (d\omega_1) \wedge \omega_2 + (-1)^k \omega_1 \wedge d\omega_2, \quad \omega_1 \in A^k(M,V),\ \omega_2 \in A^l(M,W). \tag{3.4}$$

Similarly for $F : M \to N$ a C^∞ map of C^∞ manifolds we have an induced map $F^* : A^*(N,V) \to A^*(M,V)$. Also if $P : V \to W$ is a linear map it clearly induces a map $P : A^*(M,V) \to A^*(M,W)$ commuting with d and induced maps F^* as above.

Now let G be a Lie group. The Lie algebra $\mathfrak{g}$ of G is as usual the set of left-invariant vector fields on G. This can also be identified with the tangent space of G at the unit element $1 \in G$. For $g \in G$ let $\operatorname{Ad}(g) : \mathfrak{g} \to \mathfrak{g}$ be the <u>adjoint</u>

representation, i.e., the differential at 1 of the map $x \mapsto gxg^{-1}$.

Now let $\pi : E \to M$ be a principal G-bundle. For $x \in E$ the map $G \to E$ given by $g \mapsto x \cdot g$ induces an injection $\nu_x : \mathfrak{g} \to T_x(E)$ and the quotient space is naturally identified with $T_{\pi(x)}(M)$. That is, we have an exact sequence

$$(3.5) \qquad 0 \longrightarrow \mathfrak{g} \xrightarrow{\nu_x} T_x(E) \xrightarrow{\pi_*} T_{\pi(x)}(M) \longrightarrow 0.$$

The vectors in the image of ν_x are called <u>vertical</u> and we want to single out a complement in $T_x(E)$ of <u>horizontal</u> vectors, i.e., we want to split the exact sequence (3.5). This of course is equivalent to a linear map $\theta_x : T_x(E) \to \mathfrak{g}$ such that

$$(3.6) \qquad \theta_x \circ \nu_x = \mathrm{id} \ .$$

It is therefore natural to define a connection in E simply to be a 1-form $\theta \in A^1(E,\mathfrak{g})$ such that (3.6) holds for all $x \in E$. However, we want a further condition on θ. To motivate this consider the trivial bundle $E = M \times G \to M$ and let θ be the 1-form on E given by

$$(3.7) \qquad \theta_{(x,g)} = (L_{g^{-1}} \circ \pi_2)_*, \quad x \in M,\ g \in G,$$

where $\pi_2 : M \times G \to G$ is the projection and $L_{g^{-1}} : G \to G$ is left translation by g. Now for $g \in G$ let $R_g : E \to E$ denote the map given by the action of g on the principal G-bundle E, i.e. for $E = M \times G$, by the right action on G and the trivial action on M.

<u>Lemma 3.8</u>. For θ defined by (3.7) we have

$$R_g^*\theta = \mathrm{Ad}(g^{-1}) \circ \theta, \quad \forall g \in G,$$

where $\mathrm{Ad}(g^{-1}) \circ : A^1(E,\mathfrak{g}) \to A^1(E,\mathfrak{g})$ is induced by

$\mathrm{Ad}(g^{-1}) : \mathfrak{g} \to \mathfrak{g}$.

Proof. Since θ is induced via π_2 from G it is enough to consider $M = pt$. That is, θ is the 1-form on G defined by

$$\theta_\gamma = (L_{\gamma^{-1}})_* : T_\gamma(G) \to T_1(G) = \mathfrak{g}.$$

Then

$$(R_g^*\theta)_\gamma = \theta_{\gamma g} \circ (R_g)_* = (L_{g^{-1}\gamma^{-1}})_* \circ (R_g)_*$$

$$= (L_{g^{-1}})_* \circ (L_{\gamma^{-1}})_* \circ (R_g)_* = \mathrm{Ad}(g^{-1}) \circ \theta_\gamma.$$

With this motivation we have

Definition 3.9. A connection in a principal G-bundle $\pi : E \to M$ is a 1-form $\theta \in A^1(E,\mathfrak{g})$ satisfying:

(i) $\theta_x \circ \nu_x = \mathrm{id}$ where $\nu_x : \mathfrak{g} \to T_x(E)$ is the differential of the map $g \mapsto xg$.

(ii) $R_g^*\theta = \mathrm{Ad}(g^{-1}) \circ \theta$, $\forall g \in G$, where $R_g : E \to E$ is given by the action of g on E.

Remark 1. If $H_x \subseteq T_x(E)$ is the subspace of horizontal vectors, i.e. $H_x = \ker \theta_x$, then (ii) is equivalent to

(ii)' $R_{g*}H_x = H_{xg}$, $\forall x \in E$, $\forall g \in G$.

In fact (ii) clearly implies (ii)' and since both sides of (ii) vanish on horizontal vectors (granted (ii)') it is enough to check (ii) on vertical vectors in which case (ii) is obvious from (i) and Lemma 3.8.

Remark 2. By Lemma 3.8 the product bundle $M \times G \to M$ has a connection given by (3.7). This is called the *flat* connection or the *Maurer-Cartan* connection of $M \times G$. Notice that if $\varphi : F \to E$ is an isomorphism of G-bundles and if E has a connection θ then $\varphi^*\theta$ defines a connection in F. In particular every trivial bundle has a connection induced from the flat connection in the product bundle. This is also called the *flat* connection *induced by the* given *trivialization*.

The following proposition is obvious.

Proposition 3.10. Any convex combination of connections is again a connection. More precisely: Let $\theta_1,\ldots,\theta_k$ be connections in $\pi : E \to M$ and let $\lambda_1,\ldots,\lambda_k$ be realvalued functions on M with $\sum_i \lambda_i = 1$. Then $\theta = \sum_i \lambda_i \theta_i$ is again a connection in E.

Corollary 3.11. Any principal G-bundle $\pi : E \to M$ on a paracompact manifold M has a connection.

Proof. By Remark 2 above every trivial bundle has a flat connection. In general local trivializations define flat connections θ_α in $E|U_\alpha$ for $\{U_\alpha\}_{\alpha\in\Sigma}$ a covering of M. Now choose a partition of unity $\{\lambda_\alpha\}$ and put $\theta = \sum_\alpha \lambda_\alpha \theta_\alpha$. It follows from Proposition 3.10 that θ is a connection.

Exercise 5. a) Suppose we have a bundle map of principal G-bundles

$$\begin{array}{ccc} F & \xrightarrow{\bar{f}} & E \\ \downarrow & & \downarrow \\ N & \xrightarrow{f} & M . \end{array}$$

If E has a connection θ then $\bar{f}^*\theta$ defines a connection in F.

b) If $E \to M$ is a trivial G-bundle then there is a bundle map

$$\begin{array}{ccc} E & \to & G \\ \downarrow & & \downarrow \\ M & \to & pt. \end{array}$$

and the flat connection is just the induced connection of the Maurer-Cartan connection in the G-bundle $G \to pt.$

Now consider a principal G-bundle $\pi : E \to M$ with connection θ. For $X \in T_x(E)$ a tangent vector we have already introduced the term <u>vertical</u> for $X \in \text{im}\, \nu_x$, $\nu_x : \mathfrak{g} \to T_x(E)$, and <u>horizontal</u> for $X \in H_x = \ker \theta_x$. Now suppose $\omega \in A^*(E,V)$ is a k-form with coefficients in some vectorspace V. We will say that ω is <u>horizontal</u> if $\omega(X_1,\ldots,X_k) = 0$ whenever just one of the vectors $X_1,\ldots,X_k \in T_x(E)$ is vertical. If V is a (left) representation of G then we will say that ω is <u>equivariant</u> if $R_g^*\omega = g^{-1} \cdot \omega$, $\forall g \in G$. In particular if V is the <u>trivial</u> representation an equivariant form is called <u>invariant</u>. Notice that the invariant horizontal forms on E with coefficients in $\mathbb{R}$ are exactly the forms in the image of $\pi^* : A^*(M) \to A^*(E)$. In fact suppose $\omega \in A^*(E)$ is horizontal and invariant; then we define $\overline{\omega} \in A^k(M)$ as follows: For $p \in M$ and $\overline{X}_1,\ldots,\overline{X}_k \in T_p(M)$ choose $x \in \pi^{-1}(p)$ and $X_1,\ldots,X_k \in T_x(E)$ such that $\pi_* X_i = \overline{X}_i$, $i = 1,\ldots,k$ and put

$$\overline{\omega}(\overline{X}_1,\ldots,\overline{X}_k) = \omega(X_1,\ldots,X_k).$$

This is then independent of the choices of x and $X_1,\ldots,X_k$. Furthermore if $\overline{X}_1,\ldots,\overline{X}_k$ are extended to C^∞ vector fields on M we can by local triviality of E extend $X_1,\ldots,X_k$ in a neighbourhood of x to C^∞ vector fields satisfying $\pi_* X_i = \overline{X}_i$, so $\overline{\omega}(\overline{X}_1,\ldots,\overline{X}_k)$ is C^∞ in a neighbourhood of x.

Now consider the connection from $\theta \in A^1(E,\mathfrak{g})$. Observe that θ is an equivariant 1-form with coefficients in $\mathfrak{g}$ with the adjoint action of G. Also let $[\theta,\theta] \in A^2(E,\mathfrak{g})$ denote the image of $\theta \wedge \theta$ under the map $A^2(E,\mathfrak{g}\otimes\mathfrak{g}) \to A^2(E,\mathfrak{g})$ induced by the bracket $[-,-] : \mathfrak{g}\otimes\mathfrak{g} \to \mathfrak{g}$. Then we have:

Proposition 3.12. a) Let $E = M \times G$ with the flat connection θ. Then

$$(3.13) \qquad d\theta = -\tfrac{1}{2}[\theta,\theta].$$

b) Let $\pi : E \to M$ be a principal G-bundle with connection θ and let $\Omega \in A^2(E,\mathfrak{g})$ be the curvature form defined by

$$(3.14) \qquad d\theta = -\tfrac{1}{2}[\theta,\theta] + \Omega$$

(the structural equation). Then Ω is horizontal and equivariant.

c) Furthermore Ω satisfies the Bianchi identity

$$(3.15) \qquad d\Omega = [\Omega,\theta].$$

In particular $d\Omega$ vanishes on sets of horizontal vectors.

Proof. a) follows from b) since by Exercise 5 θ is induced from the principal G-bundle $G \to pt$ and therefore $\Omega = 0$ because it is horizontal by b).

b) It is obvious that Ω is equivariant since θ and hence both $d\theta$ and $[\theta,\theta]$ are equivariant (for the second one observe that clearly $\mathrm{Ad}(g) : \mathfrak{g} \to \mathfrak{g}$ preserves the Lie bracket).

To see that Ω is horizontal we must show for $x \in E$ and for any $X,Y \in T_x(E)$ with X vertical that

$$(3.16) \qquad (d\theta)(X,Y) = -\tfrac{1}{2}[\theta,\theta](X,Y) = -\tfrac{1}{2}[\theta(X),\theta(Y)].$$

In order to show (3.16) it is enough to consider 1) Y vertical and 2) Y horizontal.

1) First notice that for any vector $A \in \mathfrak{g}$ there is an associated C^∞ vector field A^* on E defined by $A^*_x = \nu_x(A)$ where $\nu_x : \mathfrak{g} \to T_x(E)$ as usual is induced by $g \mapsto xg$. Observe that the associated 1-parameter group of diffeomorphisms is $\{R_{g_t}\}$, $t \in \mathbb{R}$, where $g_t = \exp tA$, $t \in \mathbb{R}$. Also it is easy to see that for $A, B \in \mathfrak{g}$

$$[A,B]^* = [A^*,B^*]. \tag{3.17}$$

In fact by local triviality it is enough to prove this for a trivial G-bundle $E = M \times G$ in which case $A^* = 0 \oplus \tilde{A}$ where $\tilde{A}$ is the left invariant vector field on G associated to A. Therefore (3.17) is immidiate from the definition of the Lie bracket in $\mathfrak{g}$.

Now, to prove (3.16) for X and Y vertical it is clearly enough to prove

$$(d\theta)(A^*,B^*) = -\tfrac{1}{2}[\theta(A^*),\theta(B^*)], \quad A,B \in \mathfrak{g}.$$

But since $\theta(A^*) = A$, $\theta(B^*) = B$ are constants we conclude

$$\begin{aligned}(d\theta)(A^*,B^*) &= -\tfrac{1}{2}\theta([A^*,B^*]) = -\tfrac{1}{2}\theta([A,B]^*)\\ &= -\tfrac{1}{2}[A,B] = -\tfrac{1}{2}[\theta(A^*),\theta(B^*)].\end{aligned}$$

2) Again extend X to a vector field of the form A^*, $A \in \mathfrak{g}$. Also for Y horizontal extend it to a horizontal C^∞ vector field also denoted by Y (first extend Y to any C^∞ vector field Z and then put $Y_y = Z_y - \nu_y \circ \theta_y(Z_y)$, $y \in E$). Since Y is horizontal the right hand side of (3.16) vanishes. So we must show

$$(d\theta)(A^*,Y) = 0 \quad \text{for } A \in \mathfrak{g}, \ Y \text{ a horizontal vector field.} \tag{3.18}$$

Now since $\theta(A^*) = A$ is constant and $\theta(Y) = 0$

$$(d\theta)(A^*,Y) = -\tfrac{1}{2}\theta([A^*,Y]).$$

As remarked in 1) the 1-parameter group associated to A^* is R_{g_t}, $g_t = \exp tA$, $t \in \mathbb{R}$. Therefore

$$[A^*,Y]_x = \lim_{t\to 0} \frac{1}{t}(Y_x^{g_t} - Y_x)$$

where $Y_x^{g_t} = (R_{g_t})_*(Y_{xg_t^{-1}})$. Since

$$\theta(Y_x^{g_t}) = \mathrm{Ad}(g_t^{-1}) \circ \theta(Y_{xg_t^{-1}}) = 0 \quad \text{and} \quad \theta(Y_x) = 0,$$

we conclude

$$\theta([A^*,Y]_x) = 0$$

which proves (3.18) and hence proves b).

c) Differentiating (3.14) we get

$$0 = d\Omega - \tfrac{1}{2}[d\theta,\theta] + \tfrac{1}{2}[\theta,d\theta]$$

$$= d\Omega - [d\theta,\theta] = d\Omega - [\Omega,\theta] + \tfrac{1}{2}[[\theta,\theta],\theta]$$

$$= d\Omega - [\Omega,\theta]$$

since $[[\theta,\theta],\theta] = 0$ by the Jacobi identity. This proves the proposition.

<u>Remark</u>. Let X, Y be <u>horizontal</u> vector fields on E. Then by (3.14)

$$\Omega(X,Y) = -\tfrac{1}{2}\theta([X,Y]) \tag{3.19}$$

which gives another way of defining Ω.

<u>Definition 3.20</u>. A connection θ in a principal G-bundle is called <u>flat</u> if the curvature form vanishes, that is, $\Omega = 0$.

Theorem 3.21. A connection θ in a principal G-bundle $\pi : E \to M$ is flat iff around every point of M there is a neighbourhood U and a trivialization of $E|U$ such that the restriction of θ to $E|U$ is induced from the flat connection in $U \times G$.

Proof. $\Leftarrow$ is obvious by Proposition 3.12 a).

$\Rightarrow$: Suppose $\Omega = 0$. For $x \in E$ let $H_x \subseteq T_x(E)$ be the subspace of horizontal vectors, i.e. $X \in H_x$ iff $\theta(X) = 0$. This clearly defines a distribution on E (i.e. a differentiable subbundle of $T(E)$). By (3.19) this is an integrable distribution hence by Frobenius' integrability theorem defines a foliation (see e.g. M. Spivak [29, Chapter 6]) such that H_x is the tangent space to the leaf through x. It follows from Remark 1 following Definition 3.9 that $R_g : E \to E$, $g \in G$, maps any leaf diffeomorphically onto some (possibly different) leaf of the foliation.

Now let $p \in M$ and choose $x \in \pi^{-1}(p)$ and consider the leaf F through x. Since $T_x(F) = H_x$ and since $\pi_x : H_x \to T_p(M)$ is an isomorphism we can find a neighbourhood U of p and a neighbourhood V of x in F such that $\pi : V \to U$ is a diffeomorphism. The inverse $s : U \to V$ therefore defines a section of $E|U$; hence by exercise 2 $E|U$ is trivial. In fact the trivialization is given by $\psi^{-1} : \pi^{-1}(U) \to U \times G$ where $\psi : U \times G \to \pi^{-1}(U)$ is defined by

$$\psi(q,g) = s(q) \cdot g, \quad q \in U, \ g \in G.$$

Now let θ' be the connection in $E|U$ induced from the flat connection in $U \times G$. Then it is obvious that the horizontal subspace in $T_{y \cdot g}(E)$, $y \in V$, $g \in G$, is $(R_g)_*(T_y(V)) = R_{g*}H_y = H_{yg}$, so θ and θ' defines the same horizontal subspaces and therefore must agree.

Corollary 3.22. Let $\pi : E \to M$ be a principal G-bundle. The following are equivalent:

1) E has a connection with vanishing curvature.

2) There is a covering of M by open sets $\{U_\alpha\}_{\alpha\in\Sigma}$ and a set of transition functions $\{g_{\alpha\beta}\}$ for E such that $g_{\alpha\beta} : U_\alpha \cap U_\beta \to G$ is constant for all $\alpha,\beta \in \Sigma$.

3) Let G_d be the group G with the discrete topology. Then E has a reduction to G_d.

Proof. 2) and 3) are equivalent by Exercise 4.

2) ⇒ 1): Let $\varphi_\alpha : \pi^{-1}U_\alpha \to U_\alpha \times G$, $\alpha \in \Sigma$, be the trivializations with the constant transition functions $g_{\alpha\beta}$. Let θ_α be the connection in $E|U_\alpha$ induced from the flat connection in $U_\alpha \times G$. Now there is a commutative diagram of bundle maps

$$\begin{array}{ccc}
U_\alpha \cap_\beta \times G & \xrightarrow{\varphi_\alpha \circ \varphi_\beta^{-1}} & U_\alpha \cap U_\beta \times G \\
\downarrow \pi_2 & & \downarrow \pi_2 \\
G & \xrightarrow{L_{g_{\alpha\beta}}} & G
\end{array}$$

and let θ_0 be the Maurer-Cartan connection in $G \to pt$. By definition θ_0 is left invariant and therefore

$$(\varphi_\alpha \circ \varphi_\beta^{-1})^* \pi_2^* \theta_0 = \pi_2^* \theta_0$$

or equivalently θ_α and θ_β agree on $E|U_\alpha \cap U_\beta$. Therefore we can define a global connection θ in E which agree with θ_α on $E|U_\alpha$. Clearly θ has vanishing curvature since θ_α has for all α.

1) ⇒ 2): Now let θ be a connection in E with vanishing curvature. By Theorem 3.21 we can cover M by open sets $\{U_\alpha\}$ and find trivializations $\varphi_\alpha : U \to U_\alpha \times G$ such that $\theta|\pi^{-1}U_\alpha$ is induced from the flat connection in $U_\alpha \times G$. Now fix $\alpha, \beta \in \Sigma$ and let

$$\varphi = \varphi_\alpha \circ \varphi_\beta^{-1} : U_\alpha \cap U_\beta \times G \to U_\alpha \cap U_\beta \times G.$$

Again let θ_0 be the flat connection in $U_\alpha \cap U_\beta \times G$. Then clearly $\varphi^*\theta_0 = \theta_0$ so φ permutes the leaves of the horizontal foliation, i.e., the sets of the form $(U_\alpha \cap U_\beta) \times g$, $g \in G$. In particular $\varphi(U_\alpha \cap U_\beta \times 1) = (U_\alpha \cap U_\beta) \times g_0$ for some $g_0 \in G$, and it follows that

$$\varphi(x,g) = (x, g_0 g) \quad \forall x \in U_\alpha \cap U_\beta,\ g \in G.$$

Hence the transition function $g_{\alpha\beta}$ is constantly equal to g_0.

__Exercise 6__. Let $\alpha : H \to G$ be a Lie group homomorphism and let $F \to M$ be a principal H-bundle with connection θ_F. Show that if $\varphi : F \to E$ is the extension to G then there is a connection θ_E in E such that $\varphi^*\theta_E = \alpha_* \circ \theta_F$, where α_* is the induced map of Lie algebras.

__Exercise 7__. Let M be a manifold and let $F(M) = F(TM)$ be the frame bundle of the tangent bundle, $\pi : F(M) \to M$ the projection. The structure group is $Gl(n,\mathbb{R})$ with Lie algebra $\mathfrak{gl}(n,\mathbb{R}) = \mathrm{Hom}(\mathbb{R}^n, \mathbb{R}^n)$. Since $x \in \pi^{-1}(p)$, $p \in M$, is an isomorphism $x : \mathbb{R}^n \to T_p(M)$ there is a 1-form ω on $F(M)$ with coefficients in $\mathbb{R}^n$ defined by

$$\omega_x = x^{-1} \circ \pi_*.$$

a) Show that ω on $F(M)$ is a horizontal equivariant 1-form, where $Gl(n,\mathbb{R})$ acts on $\mathbb{R}^n$ by the usual action.

b) For $M = \mathbb{R}^n$ and for $\theta \in A^1(F(M)), \mathfrak{gl}(n,\mathbb{R}))$ the connection in $F(\mathbb{R}^n)$ defined by the natural trivialization $T\mathbb{R}^n \cong \mathbb{R}^n \times \mathbb{R}^n$, show that

$$d\omega = -\theta \wedge \omega$$

where the wedge-product denotes the composite map

$$\begin{array}{ccc} A^1(F(M), \mathfrak{gl}(n,\mathbb{R})) \otimes A^1(F(M), \mathbb{R}^n) & \xrightarrow{\ \wedge\ } & A^2(F(M), \mathfrak{gl}(n,\mathbb{R}) \otimes \mathbb{R}^n) \\ & \searrow & \downarrow \\ & & A^2(F(M), \mathbb{R}^n). \end{array}$$

(Hint: Notice that $F(\mathbb{R}^n) = \mathbb{R}^n \times Gl(n,\mathbb{R}) \subseteq \mathbb{R}^n \times \mathbb{R}^{n^2}$ with coordinates $y = (y_1, \ldots, y_n) \in \mathbb{R}^n$ and $X = \{x_{ij}\}_{i,j=1,\ldots,n}$ a real $n \times n$-matrix. Then $\theta = X^{-1}dX$ and $\omega = X^{-1}dy$).

For M a general manifold and θ a connection in $F(M)$ show that the <u>torsion-form</u> $\Theta \in A^2(F(M), \mathbb{R}^n)$ defined by

$$d\omega = -\theta \wedge \omega + \Theta \tag{3.23}$$

is equivariant and horizontal.

c) With respect to the canonical basis of $\mathbb{R}^n$ we write

$$\omega = \begin{pmatrix} \omega^1 \\ \cdot \\ \cdot \\ \cdot \\ \omega^n \end{pmatrix}$$

where $\omega^1, \ldots, \omega^n$ are usual 1-forms on $F(M)$. Similarly we write

$$\theta = \begin{pmatrix} \theta^1_1 & \cdots & \theta^1_n \\ \vdots & & \vdots \\ \theta^n_1 & \cdots & \theta^n_n \end{pmatrix}, \quad \Theta = \begin{pmatrix} \Theta^1 \\ \vdots \\ \Theta^n \end{pmatrix}$$

Then (3.23) takes the form

(3.23)' $$d\omega^i = -\sum_j \theta^i_j \wedge \omega^j + \Theta^i, \quad i = 0,\dots,n.$$

d) Show that every horizontal 1-form α on $F(M)$ is of the form $\alpha = \sum_i f_i \omega^i$, where f_i are real valued C^∞ functions on $F(M)$.

e) Now suppose M is given a Riemannian metric and let θ be a connection in the orthogonal frame bundle $F_O(M)$. Let ω and Θ be defined on $F_O(M)$ exactly as for $F(M)$ above. Show that (3.23) still holds and that on $F_O(M)$

(3.24) $$\theta^i_j = -\theta^j_i, \quad i,j = 1,\dots,n.$$

Furthermore show that if $\Theta = 0$ then θ is uniquely determined by (3.23) and (3.24). (Hint: Show first that if $\alpha = (\alpha_j)$ is a row of horizontal 1-forms satisfying $\sum_j \alpha_j \wedge \omega^j = 0$ and if we write $\alpha_j = \sum_i f_{ij}\omega^i$ as in d), then $f_{ij} = f_{ji}$).

f) Conclude that for every Riemannian manifold M the framebundle $F_O(M)$ has a unique torsion free connection (the <u>Levi-Civita connection</u>). Notice that by Exercise 6 this extends to a well-defined connection in $F(M)$.

<u>Exercise 8</u>. Let M be a manifold and $V \to M$ an n-dimensional vector bundle. Let $\pi : F \to M$ be the associated principal $Gl(n,\mathbb{R})$-bundle, i.e. the bundle of n-frames in V. Again $\mathfrak{gl}(n,\mathbb{R}) = \mathrm{Hom}(\mathbb{R}^n, \mathbb{R}^n)$ is the Lie algebra of $Gl(n,\mathbb{R})$.

a) Show that for $\theta \in A^1(F,\mathfrak{gl}(n,\mathbb{R}))$, θ a connection in F, (3.14) takes the form

(3.25) $$d\theta = -\theta \wedge \theta + \Omega$$

where the wedge-product denotes the composite

$$\begin{array}{ccc} A^1(F,\mathfrak{gl}(n)) \otimes A^1(F,\mathfrak{gl}(n)) & \xrightarrow{\ \wedge\ } & A^2(F,\mathfrak{gl}(n) \otimes \mathfrak{gl}(n)) \\ & \searrow & \downarrow \\ & & A^2(F,\mathfrak{gl}(n)) \end{array}$$

(here $\mathfrak{gl}(n,\mathbb{R}) \otimes \mathfrak{gl}(n,\mathbb{R}) \to \mathfrak{gl}(n,\mathbb{R})$ is given by composition of maps of $\mathbb{R}^n$ into $\mathbb{R}^n$). Furthermore, with respect to the canonical basis of $\mathfrak{gl}(n,\mathbb{R})$, θ and Ω are given by matrices

$$\begin{pmatrix} \theta^1_1 & \cdots & \theta^1_n \\ \vdots & & \vdots \\ \theta^n_1 & \cdots & \theta^n_n \end{pmatrix}, \quad \begin{pmatrix} \Omega^1_1 & \cdots & \Omega^1_n \\ \vdots & & \vdots \\ \Omega^n_1 & \cdots & \Omega^n_n \end{pmatrix}$$

of 1- and 2-forms respectively.

Show that (3.25) is equivalent to

(3.25)' $$d\theta^i_j = -\sum_k \theta^i_k \wedge \theta^k_j + \Omega^i_j, \qquad i,j = 1,\ldots,n.$$

b) Observe that C^∞ sections of V are in 1-1 correspondence with equivariant C^∞ functions of F into $\mathbb{R}^n$ where $Gl(n,\mathbb{R})$ acts on $\mathbb{R}^n$ in the usual way. The set of C^∞ sections of V is denoted $\Gamma(V)$.

Similarly show that C^∞ sections of $T^*M \otimes V$ are in 1-1 correspondence with equivariant horizontal 1-forms on F with coefficients in $\mathbb{R}^n$. Alternatively $\ell \in \Gamma(T^*M \otimes V)$ associates to every vector $X_p \in T_p(M)$ an element $\ell_{X_p} \in V_p$ such that

(i) $\ell_{X_p+Y_p} = \ell_{X_p} + \ell_{Y_p}$, $\ell_{\lambda X_p} = \lambda \ell_{X_p}$, $\lambda \in \mathbb{R}$,

(ii) if X is a C^∞ vector field on M then the function $p \mapsto \ell_{X_p}$ is a C^∞ section of V.

c) Let again θ be a connection in F. For any $s \in \Gamma(V)$ define $\nabla(s) \in A^1(F, \mathbb{R}^n)$ by

$$ds = -\theta \cdot s + \nabla(s) \tag{3.26}$$

(here s is considered as a function of F into $\mathbb{R}^n$). Show that $\nabla(s)$ is horizontal and equivariant, hence defines $\nabla(s) \in \Gamma(T^*M \otimes V)$.

d) For $s \in \Gamma(V)$ and $X_p \in T_p(M)$ let $\nabla(s) \in \Gamma(T^*M \otimes V)$ as in c) and let $\nabla_{X_p}(s) = \nabla(s)_{X_p} \in V_p$ as defined in b). This is called the <u>covariant derivative</u> of s in the direction X_p and ∇ is called the <u>covariant differential</u> corresponding to θ. Show that ∇ satisfies:

(i) $\nabla_{X_p+Y_p}(s) = \nabla_{X_p}(s) + \nabla_{Y_p}(s)$, $\nabla_{\lambda X_p}(s) = \lambda\nabla_{X_p}(s)$,

$s \in \Gamma(M)$, $\lambda \in \mathbb{R}$.

(ii) If X is a C^∞ vector field on M then the function $p \mapsto \nabla_{X_p}(s)$ is a C^∞ section of V. This is denoted $\nabla_X(s)$.

(iii) $\nabla_X(fs) = X(f)s + f\nabla_X(s)$ for $s \in \Gamma(V)$, f a C^∞ real valued function on M and $X(f)$ the directional derivative of f.

e) As before let θ be a connection in $\pi : F \to M$. Show that for $\gamma : [a,b] \to M$ a C^∞ curve and $x \in \pi^{-1}(\gamma(a))$ there is a unique liftet curve $\bar{\gamma} : [a,b] \to F$ with $\bar{\gamma}(a) = x$, $\pi \circ \bar{\gamma} = \gamma$, such that the tangents of $\bar{\gamma}$ are all horizontal. Notice that this lift defines an isomorphism (the "parallel translation along γ") $\tau_{\gamma(t)} : V_{\gamma(a)} \to V_{\gamma(t)}$, $t \in [a,b]$.

f) For $X_p \in T_p(M)$ let $\gamma : [-\varepsilon,\varepsilon] \to M$, $\varepsilon > 0$ be a C^∞ curve with $\gamma(0) = p$, $\gamma'(0) = X_p$. Let $\tau_t : V_p \to V_{\gamma(t)}$ be parallel translation along γ. Show that for $s \in \Gamma(V)$

$$(3.27) \qquad \nabla_{X_p}(s) = \lim_{t\to 0} \frac{\tau_t^{-1} s(\gamma(t)) - s(p)}{t} .$$

(Hint: Observe that in some neighbourhood U of p there is a section v of $F|U$ such that $v \circ \gamma$ defines a <u>horizontal</u> lift of γ. Now write $s = \sum_i a_i v_i$ where $(v_1,\ldots,v_n)$ are the components of v and $a_i : U \to \mathbb{R}$, $i = 1,\ldots,n$, are C^∞ functions).

g) Now let $\Omega \in A^2(F,\mathfrak{gl}(n,\mathbb{R}))$ be the curvature form of θ. Show that for any $s \in \Gamma(V)$, interpreted as an equivariant function of F into $\mathbb{R}^n$, we have

$$(3.28) \qquad d\nabla(s) = \Omega \cdot s - \theta \wedge \nabla(s) .$$

Notice that for X and Y vector fields on M Ω defines a section $\Omega(X,Y) \in \Gamma(\mathrm{Hom}(V,V))$. Show that

$$(3.29) \qquad \Omega(X,Y)(s) = \tfrac{1}{2}(\nabla_X \circ \nabla_Y - \nabla_Y \circ \nabla_X - \nabla_{[X,Y]})(s), \ \forall s \in \Gamma(V).$$

h) Now let $V = TM$ and let ω be the 1-form considered in Exercise 7. Let θ be a connection in $F(M)$ with torsion form Θ. Observe that for X, Y vector fields on M Θ defines a section of TM, that is, a new vector field $\Theta(X,Y)$ and show that this is given by

$$(3.30) \qquad \Theta(X,Y) = \tfrac{1}{2}(\nabla_X(Y) - \nabla_Y(X) - [X,Y])$$

where ∇ is defined in d).

(Hint: Notice first that for any vector field $\tilde{X}$ on $F(M)$

which is a lift of a vector field X on M (that is, $\pi_* \tilde{X}_x = X_{\pi x}$, $\forall x \in F(M)$) the function $\omega(\tilde{X}) : F(M) \to \mathbb{R}^n$ is the equivariant function corresponding to X as in b) above.

Note. Our treatment of principal bundles and connections follows closely the exposition by S. Kobayashi and K. Nomizu [17, Chapter I and II].

4. The Chern-Weil homomorphism

We now come to the main object of these lectures, namely to construct characteristic cohomology classes for principal G-bundles by means of a connection. First some notation:

Let V be a finite dimensional real vector space. For $k \geqq 1$ let $S^k(V^*)$ denote the vector space of symmetric multilinear real valued functions in k variables on V. Equivalently $P \in S^k(V^*)$ is a linear map $P : V \otimes \ldots \otimes V \to \mathbb{R}$ which is invariant under the action of the symmetric group acting on $V \otimes \ldots \otimes V$. There is a product

$$\circ : S^k(V^*) \otimes S^l(V^*) \to S^{l+k}(V^*)$$

defined by

(4.1)
$$P \circ Q(v_1,\ldots,v_{k+l}) = \\ = \frac{1}{(k+l)!} \sum_\sigma P(v_{\sigma 1},\ldots,v_{\sigma k}) \cdot Q(v_{\sigma(k+1)},\ldots,v_{\sigma(k+l)})$$

where σ runs through all permutations of $1,\ldots,k+l$. Let $S^*(V^*) = \coprod_{k \geqq 0} S^k(V^*)$ $(S^0(V^*) = \mathbb{R})$; then $S^*(V^*)$ is a graded algebra.

Exercise 1. Let $\{e_1,\ldots,e_n\}$ be a basis for V and let $\mathbb{R}[x_1,\ldots,x_n]^k$ be the set of homogeneous polynomials of degree k in some variables $x_1,\ldots,x_n$. Show that the mapping

$$\sim : S^k(V^*) \to \mathbb{R}[x_1,\ldots,x_n]^k$$

defined by

$$\tilde{P}(x_1,\ldots,x_n) = P(v,\ldots,v), \quad v = \sum_i x_i e_i,$$

for $P \in S^k(V^*)$, is an isomorphism and that $\sim : S^*(V^*) \to \mathbb{R}[x_1,\dots,x_n]$ is an algebra isomorphism. This shows that P is determined by the <u>polynomial</u> <u>function</u> on V given by $v \mapsto P(v,\dots,v)$. The inverse of $\sim$ is called <u>polarization</u>.

Now let G be a Lie group with Lie algebra $\mathfrak{g}$. Then the adjoint representation induces an action of G on $S^k(\mathfrak{g}^*)$ for every k:

$$(gP)(v_1,\dots,v_k) = P(\mathrm{Ad}(g^{-1})v_1,\dots,\mathrm{Ad}(g^{-1})v_k),$$
$$v_1,\dots,v_k \in \mathfrak{g}, \quad g \in G.$$

Let $I^k(G)$ be the G-invariant part of $S^k(\mathfrak{g}^*)$. Notice that the multiplication (4.1) induces a multiplication

$$(4.2) \qquad I^k(G) \otimes I^l(G) \to I^{k+l}(G).$$

In view of Exercise 1 $I^*(G)$ is called the algebra of <u>invariant</u> <u>polynomials</u> on $\mathfrak{g}$.

Now consider a principal G-bundle $\pi : E \to M$ on a differentiable manifold M, and suppose θ is a connection in E with curvature form $\Omega \in A^2(E,\mathfrak{g})$. Then for $k \geq 1$ we have

$$\Omega^k = \Omega \wedge \dots \wedge \Omega \in A^{2k}(E,\mathfrak{g}\otimes\dots\otimes\mathfrak{g}) = A^{2k}(E,\mathfrak{g}^{\otimes k})$$

so $P \in I^k(G)$ gives rise to a 2k-form $P(\Omega^k) \in A^{2k}(E)$. Since Ω is horizontal also $P(\Omega^k)$ is horizontal, and since Ω is equivariant and P <u>invariant</u> $P(\Omega^k)$ is an invariant horizontal 2k-form. Hence $P(\Omega^k)$ is the lift of a 2k-form on M which we also denote by $P(\Omega^k)$.

<u>Theorem 4.3</u>. a) $P(\Omega^k) \in A^{2k}(M)$ is a closed form.

Let $w_E(P) \in H^{2k}(A^*(M))$ be the corresponding cohomology class. Then

b) $\omega_E(P)$ does not depend on the choice of connection and in particular does only depend on the isomorphism class of E.

c) $\omega_E : I^*(G) \to H(A^*(M))$ is an algebra homomorphism.

d) For $f : N \to M$ a differentiable map

$$\omega_{f^*E} = f^* \circ \omega_E .$$

Remark. The map ω_E is called the Chern-Weil homomorphism. Sometimes we shall just denote it by ω when the bundle in question is clear from the context. For $P \in I^*(G)$ $\omega_E(P)$ is called the characteristic class of E corresponding to P.

Proof of Theorem 4.3. a) Since $\pi^* : A^*(M) \to A^*(E)$ is injective it is enough to show that $dP(\Omega^k) = 0$ in $A^*(E)$. Now since P is symmetric and Ω a 2-form

$$dP(\Omega^k) = kP(d\Omega \wedge \Omega^{k-1}) = kP([\Omega,\theta] \wedge \Omega^{k-1}) \tag{4.4}$$

by (3.15). On the other hand since $P \in S^k(\mathfrak{g}^*)$ is invariant we have

$$P(\mathrm{Ad}(g_t)Y_1,\ldots,\mathrm{Ad}(g_t)Y_k) = P(Y_1,\ldots,Y_k), \tag{4.5}$$
$$g_t = \exp tY_0, \quad Y_0,Y_1,\ldots,Y_k \in \mathfrak{g}, \quad t \in \mathbb{R}.$$

Differentiating (4.5) at $t = 0$ we get

$$\sum_{i=1}^{k} P(Y_1,\ldots,[Y_0,Y_i],\ldots,Y_k) = 0$$

or equivalently

$$\sum_{i=1}^{k} P([Y_0,Y_i],Y_1,\ldots,\hat{Y}_i,\ldots,Y_k) = 0, \quad Y_0,\ldots,Y_k \in \mathfrak{g}.$$

From this it follows that $P([\theta,\Omega] \wedge \Omega \wedge \ldots \wedge \Omega) = 0$ which together with (4.4) ends the proof of a).

b) For this we need the following easy lemma (compare Chapter 1, Exercise 5 or Lemma 1.2):

Lemma 4.6. Let $h : A^k(M \times [0,1]) \to A^{k-1}(M)$, $k = 0,1,\ldots,$ be the operator sending $\omega = ds \wedge \alpha + \beta$ to

$$h(\omega) = \int_{s=0}^{1} \alpha \qquad (h\omega = 0 \quad \text{for} \quad \omega \in A^0).$$

Then

$$\text{(4.7)} \qquad dh(\omega) + h(d\omega) = i_1^*\omega - i_0^*\omega, \qquad \omega \in A^*(M \times [0,1])$$

where $i_0(p) = (p,0)$, $i_1(p) = (p,1)$, $p \in M$.

Now suppose θ_0 and θ_1 are two connections in E with curvature forms Ω_0 and Ω_1 respectively. Consider the principal G-bundle $E \times [0,1] \to M \times [0,1]$ and let $\tilde{\theta} \in A^1(E \times [0,1])$ be the form given by

$$\tilde{\theta}_{(x,s)} = (1-s)\theta_{0x} + s\theta_{1x}, \quad (x,s) \in E \times [0,1].$$

By Proposition 3.10 $\tilde{\theta}$ is a connection in $E \times [0,1]$. Let $\tilde{\Omega}$ be the curvature form of $\tilde{\theta}$. Since $i_0^*\tilde{\theta} = \theta_0$, $i_1^*\tilde{\theta} = \theta_1$ it is obvious that $i_0^*\tilde{\Omega} = \Omega_0$ and $i_1^*\tilde{\Omega} = \Omega_1$. Now for $P \in I^k(G)$, $P(\tilde{\Omega}^k)$ is a closed 2k-form on $E \times [0,1]$ by a) above. Therefore by (4.7)

$$\begin{aligned} d(h(P(\tilde{\Omega}^k))) &= i_1^*P(\tilde{\Omega}^k) - i_0^*P(\tilde{\Omega}^k) \\ &= P(\Omega_1^k) - P(\Omega_0^k) \end{aligned}$$

and hence $P(\Omega_0^k)$ and $P(\Omega_1^k)$ represent the same cohomology class in $H^{2k}(A^*(M))$. This shows that $w_E(P)$ does not depend on the choice of connection. The second statement is obvious from this.

c) For $P \in I^1(G)$ and $Q \in I^k(G)$ it is straight forward to verify that

$$(4.8) \qquad (P \circ Q)(\Omega^{k+1}) = P(\Omega^1) \wedge Q(\Omega^k)$$

from which c) trivially follows.

d) If θ is a connection in $E \to M$ with curvature form Ω then clearly $\overline{f}^*\theta$ is a connection in $f^*E \to N$ with curvature form $\overline{f}^*\Omega$. Therefore since

$$(4.9) \qquad \overline{f}^*P(\Omega^k) = P(\overline{f}^*\Omega)^k$$

d) clearly follows.

Remark. Let $I^*_{\mathbb{C}}(G)$ be the algebra of complex valued G-invariant polynomials on $\mathfrak{g}$. Then for any principal G-bundle E with connection θ we get a similar complex Chern-Weil homomorphism

$$(4.10) \qquad I^*_{\mathbb{C}}(G) \to H(A^*(M,\mathbb{C})) \simeq H^*(M,\mathbb{C}).$$

Let us end this chapter with some examples of invariant polynomials for some classical groups. In all the examples we exhibit the polynomial function $v \mapsto P(v,\ldots,v)$, $v \in \mathfrak{g}$, for $P \in I^k(G)$.

Example 1. $G = Gl(n,\mathbb{R})$, the group of non-singular $n \times n$ matrices. The Lie algebra $\mathfrak{g} = \mathfrak{gl}(n,\mathbb{R}) = Hom(\mathbb{R}^n,\mathbb{R}^n)$ is the Lie algebra of all matrices with Lie bracket $[A,B] = AB - BA$. For $g \in G$, $Ad(g)(A) = gAg^{-1}$, for all $A \in \mathfrak{gl}(n,\mathbb{R})$. For k a positive integer let $P_{k/2}$ be the homogeneous polynomial of degree k which is the coefficient of λ^{n-k} in the polynomial in λ

(4.11) $\det(\lambda\cdot 1 - \frac{1}{2\pi}A) = \sum_k P_{k/2}(A,\ldots,A)\lambda^{n-k}, \quad A \in \mathfrak{gl}(n,\mathbb{R}).$

Clearly $P_{k/2} \in I^k(Gl(n,\mathbb{R}))$; $P_{k/2}$ is called the k/2-th <u>Pontrjagin polynomial</u>, and the Chern-Weil images are called the <u>Pontrjagin classes</u>.

<u>Example 2</u>. $G = O(n) \subseteqq Gl(N,\mathbb{R})$, the subgroup of matrices satisfying $g\,{}^tg = 1$ where ${}^tg$ is the transpose of g. The Lie algebra of $O(n)$ is $\mathfrak{o}(n) \subseteqq \mathfrak{gl}(n,\mathbb{R})$ of skew-symmetric matrices. Since for $A \in \mathfrak{o}(n)$

$$\det(\lambda 1 - \frac{1}{2\pi}A) = \det(\lambda 1 + \frac{1}{2\pi}A)$$

it follows that for k odd the restriction of $P_{k/2}$ to $\mathfrak{o}(n)$ is zero. Therefore we only consider $P_l \in I^{2l}(O(n))$, $l = 0,1,\ldots,[\frac{n}{2}]$. Notice that since every $Gl(n,\mathbb{R})$-bundle has a reduction to $O(n)$, the Chern-Weil image of $P_{k/2}$ for k odd is zero for any $Gl(n,\mathbb{R})$-bundle although the polynomials are non-zero on $\mathfrak{gl}(n,\mathbb{R})$.

<u>Example 3</u>. $G = SO(n) \subseteqq O(n)$, the subgroup of orthogonal matrices satisfying $\det(g) = 1$. The Lie algebra $\mathfrak{so}(n) = \mathfrak{o}(n)$ so again we have the Pontrjagin polynomials $P_l \in I^{2l}(SO(n))$, $l = 0,1,\ldots,[\frac{n}{2}]$.

Now suppose n is even, $n = 2m$, and consider the homogeneous polynomial Pf (for Pfaffian) of degree m given by

(4.12) $\mathrm{Pf}(A,\ldots,A) = \frac{1}{2^{2m}\pi^m m!} \sum_\sigma (\operatorname{sgn}\sigma)\, a_{\sigma 1\sigma 2}\cdots a_{\sigma(2m-1)\sigma(2m)}$

where the sum is over all permutations of $1,2,\ldots,2m$, and where $A = \{a_{ij}\}$ satisfies $a_{ij} = -a_{ji}$.

In order to see that Pf is invariant first notice that if $g = \{x_{ij}\} \in SO(n)$ then

$$gAg^{-1} = gA\,{}^tg = A'$$

where $A' = \{a'_{ij}\}$ is given by

$$a'_{ij} = \sum_{k_1,k_2} x_{ik_1} a_{k_1k_2} x_{jk_2}$$

so

$$\mathrm{Pf}(A',\ldots,A') = \sum_{k_1,\ldots,k_{2m}} a_{k_1k_2}\cdots a_{k_{2m-1}k_{2m}} \cdot$$

$$\cdot \sum_{\sigma} \mathrm{sgn}(\sigma)\, x_{\sigma 1 k_1} x_{\sigma 2 k_2} \cdots x_{\sigma(2m-1)k_{2m-1}} \cdot$$

$$\cdot\, x_{\sigma(2m)k_{2m}} .$$

The coefficient of $a_{k_1k_2}\cdots a_{k_{2m-1}k_{2m}}$ is the determinant of the matrix $\{x_{ik_j}\}$. This determinant is zero unless $(k_1,\ldots,k_{2m})$ is a permutation of $1\ldots 2m$ in which case it is the sign of the permutation since $\det\{x_{ij}\} = 1$. Hence $\mathrm{Pf}(A',\ldots,A') = \mathrm{Pf}(A,\ldots,A)$ so Pf is an invariant polynomial. Notice that if $\det\{x_{ij}\} = -1$ then

$$\mathrm{Pf}(gAg^{-1},\ldots,gAg^{-1}) = -\mathrm{Pf}(A,\ldots,A)$$

so Pf is <u>not</u> an invariant polynomial for $O(n)$. We shall later show that the Chern-Weil image of Pf is the Euler class; this is the content of the classical Gauss-Bonnet theorem.

<u>Example 4.</u> $G = Gl(n,\mathbb{C})$ has Lie algebra $\mathfrak{gl}(n,\mathbb{C}) = \mathrm{Hom}(\mathbb{C}^n,\mathbb{C}^n)$. Here we consider the <u>complex</u> valued invariant polynomials C_k which are the coefficients to λ^{n-k} in the polynomial

$$\det(\lambda\cdot 1 - \frac{1}{2\pi i} A) = \sum_k C_k(A,\ldots,A)\lambda^{n-k} \tag{4.13}$$

where A is an $n \times n$ matrix of complex numbers and $i = \sqrt{-1}$. The Chern-Weil image of these polynomials give characteristic classes with complex coefficients and they are called the Chern classes. Notice that the restriction of C_k to $\mathfrak{gl}(n,\mathbb{R})$ satisfy

$$i^k C_k(A,\ldots,A) = P_{k/2}(A,\ldots,A), \quad A \in \mathfrak{gl}(n,\mathbb{R}). \tag{4.14}$$

It follows that the l-th Pontrjagin class of a $Gl(n,\mathbb{R})$-bundle is $(-1)^l$ times the $2l$-th Chern class of the complexification. (The complexification of a principal $Gl(n,\mathbb{R})$-bundle is the extension to the group $Gl(n,\mathbb{C})$).

Example 5. $G = U(n) \subseteq Gl(n,\mathbb{C})$ is the subgroup of matrices g such that $g\,{}^t\bar{g} = 1$ ($\bar{g}$ is the complex conjugate of g). The Lie algebra is $\mathfrak{u}(n) \subseteq \mathfrak{gl}(n,\mathbb{C})$, the subalgebra of skew-hermitian matrices, that is, $A \in \mathfrak{u}(n)$ satisfy $A = -{}^t\bar{A}$. Therefore

$$\begin{aligned}\det(\lambda\cdot 1 - \frac{1}{2\pi i} A) &= \det(\lambda\cdot 1 + \frac{1}{2\pi i}\,{}^t\bar{A}) \\ &= \overline{\det(\lambda\cdot 1 - \frac{1}{2\pi i} A)}, \quad A \in \mathfrak{u}(n)\end{aligned}$$

hence the polynomials C_k defined by (4.13) are real valued when restricted to $\mathfrak{u}(n)$. The Chern-Weil image therefore lies naturally in real cohomology again.

Exercise 2. Let V be a finite dimensional vector space. Let

$$T^*(V) = \coprod_{k\geq 0} V^{\otimes k}$$

be the tensor algebra of V, i.e. the graded algebra with

$T^k(V) = V \otimes \ldots \otimes V$ (k factors) and with the natural product

$$T^k(V) \otimes T^l(V) \to T^{k+l}(V).$$

The <u>symmetric</u> algebra of V is the quotient

$$S^*(V) = T^*(V)/I$$

where I is the ideal generated by all elements of the form $v \otimes w - w \otimes v$. The image of $T^k(V)$ in $S^*(V)$ is denoted $S^k(V)$ and is called the k-th <u>symmetric power</u> of V.

a) Show that if V^* is the dual vectorspace of V then $S^k(V^*)$ is naturally isomorphic to $S^k(V^*)$, the vectorspace of symmetric multilinear forms in k variables.

b) Show that for vectorspaces V, W

$$S^k(V \otimes W) \simeq \coprod_{i+j=k} S^i(V) \otimes S^j(W).$$

<u>Exercise 3</u>. (S.-S. Chern and J. Simons [9]). Let $\pi : E \to M$ be a principal G-bundle with connection θ.

a) Show that for $P \in I^k(G)$ there is a "canonical" (2k-1)-form $TP(\theta)$ on E such that

$$dTP(\theta) = P(\Omega^k). \tag{4.15}$$

(Hint: Observe that π^*E has two connections: $\theta_1 = \bar{\pi}^*\theta$ (where $\bar{\pi} : \pi^*E \to E$ is the map of total spaces) and θ_0 the flat connection induced from the canonical trivialization of π^*E).

b) Suppose $f : N \to M$ is covered by $\bar{f} : f^*E \to E$. Then

$$TP(\bar{f}^*\theta) = \bar{f}^*TP(\theta). \tag{4.16}$$

c) Show that $TP(\theta)$ is given on E by

$$(4.17) \qquad TP(\theta) = k \int_{s=0}^{1} P(\theta \wedge \varphi_s^{k-1})$$

where $\varphi_s = s\Omega + \frac{1}{2}(s^2 - s)[\theta,\theta]$.

Exercise 4. Let $\alpha : H \to G$ be a Lie group homomorphism and let $\alpha_* : \mathfrak{h} \to \mathfrak{g}$ be the associated Lie algebra homomorphism.

a) Show that α_* induces a map $\alpha^* : I^*(G) \to I^*(H)$ defined by

$$\alpha^* P(v_1,\ldots,v_k) = P(\alpha_* v_1,\ldots,\alpha_* v_k)$$
$$v_1,\ldots,v_k \in \mathfrak{g}, \quad P \in I^k(G).$$

b) Suppose $\zeta : F \to M$ is an H-bundle with G-extension $\xi : E \to M$. Show that for $P \in I^*(G)$

$$(4.18) \qquad w_F(\alpha^* P) = w_E(P).$$

Note. Our exposition of the Chern-Weil construction follows the one by S. Kobayashi and K. Nomizu [17, Chapter XII].

5. Topological bundles and classifying spaces

In this section G denotes a Lie group as before. The notion of a topological principal G-bundle $\pi : E \to X$ on a topological space X is defined exactly as in Definition 3.1, only the words "differentiable" and "diffeomorphism" are replaced by "continuous" and "homeomorphism". The purpose of this and the following section is to show that the Chern-Weil homomorphism defines characteristic classes of topological G-bundles and in particular, the characteristic classes of differentiable bundles, as defined in the previous chapter, only depend on the underlying topological G-bundle. In this section we shall study characteristic classes from a general point of view. In the following H^* denotes cohomology with coefficients in a fixed ring Λ which is assumed to be a principal ideal domain (we shall mainly take $\Lambda = \mathbb{R}$).

Definition 5.1. A characteristic class c for principal G-bundles associates to every isomorphism class of topological principal G-bundles $\pi : E \to X$ a cohomology class $c(E) \in H^*(X)$, such that for every continuous map $f : Y \to X$ and for $\pi : E \to X$ a G-bundle

(5.2) $$c(f^*(E)) = f^*c(E).$$

We shall show that there is a topological space BG, called the classifying space for G such that the characteristic classes are in 1-1 correspondence with the cohomology classes in $H^*(BG)$. The construction is as follows:

As usual $\Delta^n \subseteq \mathbb{R}^{n+1}$ is the standard n-simplex with barycentric coordinates $t = (t_0, \ldots, t_n)$. Let $G^{n+1} = G \times \ldots \times G$ ($n+1$ times) and let

$$EG = \coprod_{n \geq 0} \Delta^n \times G^{n+1}/\sim$$

with the following idenfitications:

$$(\varepsilon^i t, (g_0, \ldots, g_n)) \sim (t, (g_0, \ldots, \hat{g}_i, \ldots, g_n)),$$
$$t \in \Delta^{n-1},\ g_0, \ldots, g_n \in G,\ i = 0, \ldots, n.$$

Now G acts on the right on EG by the action

$$(t, (g_0, \ldots, g_n))g = (t, (g_0 g, \ldots, g_n g))$$

and we let $BG = EG/G$ with $\gamma_G : EG \to BG$ the projection.

Proposition 5.3. $\gamma_G : EG \to BG$ is a principal G-bundle.

Proof. First notice that the action of G on EG is free (i.e. $xg = x \Rightarrow g = 1$), and it is easy to see that furthermore the action is strongly free in the following sense:

Let F be a space with a free G-action $F \times G \to F$ and let $F^* \subseteq F \times F$ be the set of pairs (x,y) with x and y in the same orbit. Then there is a natural map $\tau : F^* \to G$ defined by $y = x\,\tau(x,y)$ and the action is said to be strongly free if $\tau : F^* \to G$ is continuous. The following lemma is easy (compare Exercise 2 of Chapter 3):

Lemma 5.4. Let F be a space with a strongly free G-action. Then $\pi : F \to F/G$ is a trivial G-bundle iff π has a continuous section.

It follows that in order to show Proposition 5.3 it is enough to construct local sections of $\gamma_G : EG \to BG$. Equivalently, for any point $x \in EG$ we shall find a G-invariant open neighbourhood U of x and a continuous map $h : U \to G$ which is <u>equivariant</u> with respect to the right G-action on G (then the map $y \mapsto yh(y)^{-1}$ defines a section of $\gamma_G : U \to U/G$).

For this we shall use that since G is a manifold it is an <u>absolute neighbourhood retract</u> (ANR), i.e. whenever $A \subseteq X$ is a closed subspace of a <u>normal</u> space X and $f : A \to G$ is continuous, there is an extension of f to a neighbourhood of A in X. In fact any manifold M embeds in a Euclidean space $M \subseteq \mathbb{R}^q$ such that there is a neighbourhood N of M in $\mathbb{R}^q$ with a retraction $r : N \to M$ $(r|M = id)$. Hence whenever $A \subseteq X$ as above and $f : A \to M$ is continuous, there is an extension $F : X \to \mathbb{R}^q$ (by Titze's extension theorem) and then $r \circ F : F^{-1}(N) \to M$ extends f to the neighbourhood $F^{-1}(N)$.

Now let $x \in EG$ and we shall construct U and h as required above by constructing successively the restrictions to $EG(n) \subseteq EG$, where $EG(n)$ is the image of $\coprod_{k \leq n} \Delta^k \times G^{k+1}$ in EG.

First let n_0 be the smallest integer such that x is represented in $\Delta^{n_0} \times G^{n_0+1}$ by

$$x = ((t_0,\ldots,t_{n_0}),(g_0,\ldots,g_{n_0})).$$

Then all $t_0,\ldots,t_{n_0} > 0$ and we can clearly find an open neighbourhood V of $(t_0,\ldots,t_{n_0})$ such that $\overline{V} \subseteq \mathrm{int}(\Delta^{n_0})$. Define

$$U_{n_0} = V \times G^{n_0+1} \subseteq EG(n_0)$$

and let $h_{n_0} : \overline{U}_{n_0} \to G$ be the map which project onto the first coordinate of G^{n_0+1}.

Now let $n > n_0$ and suppose we have defined an invariant

open set $U_{n-1} \subseteq EG(n-1)$ and an equivariant map $h_{n-1} : \bar{U}_{n-1} \to G$. Let $p : \Delta^n \times G^{n+1} \to EG(n)$ be the natural projection and observe that p maps $\partial\Delta^n \times G^{n+1}$ into $EG(n-1)$. Let $W \subseteq \partial\Delta^n \times G^{n+1}$ be the closed subset $W = p^{-1}(\bar{U}_{n-1})$. Then since G is an ANR the map $h_{n-1} \circ p : W \to G$ extends to a map $h' : W' \to G$ where $W' \subseteq \Delta^n \times G^{n+1}$ is an open neighbourhood of W. Shrinking W' a little we can assume h' defined on $\bar{W}'$. Now consider $W'' \subseteq \Delta^n \times G^{n+1}$ defined by

$$W'' = \{(t,(g_0,\ldots,g_n)) \mid (t,(1,g_1g_0^{-1},\ldots,g_ng_0^{-1})) \in W'\}.$$

Clearly W'' is an open G-invariant set and notice that $W \subseteq W''$ since $\bar{U}_n$ and hence W is G-invariant. On the other hand we can find a G-invariant open subset $W''' \subseteq \Delta^n \times G^{n+1}$ such that

$$W''' \cap (\partial\Delta^n \times G^{n+1}) = p^{-1}(U_{n-1})$$

since $p^{-1}(U_{n-1}) \subseteq \partial\Delta^n \times G^{n+1}$ is an open G-invariant subset. Now let $U' = W'' \cap W'''$ and define $h'' : \bar{U}' \to G$ by

$$h''(t,(g_0,\ldots,g_n)) = h'(t,(1,g_1g_0^{-1},\ldots,g_ng_0^{-1}))\cdot g_0.$$

Clearly h'' extends $h_{n-1} \circ p : W \to G$ and is equivariant. $U_n = U_{n-1} \cup p(U')$ is an open invariant set in $EG(n)$ and clearly h'' and h_{n-1} defines an equivariant extension $h_n : \bar{U}_n \to G$. This construct U_n and h_n inductively, so let $U = \bigcup_n U_n$ and $h = \bigcup_n h_n$. This ends the proof of the proposition.

We can now state the main result of this chapter:

<u>Theorem 5.5</u>. The map associating to a characteristic class c for principal G-bundles the element $c(E(G)) \in H^*(BG)$ is a 1-1 correspondence.

For the proof we shall study EG and BG from a "simplicial" point of view:

Let $X = \{X_q\}$, $q = 0,1,\ldots,$ be a simplicial set and suppose that each X_q is a topological space such that all face and degeneracy operators are continuous. Then X is called a simplicial space and associated to this is the so-called fat realization, the space $\| X \|$ given by

$$\| X \| = \coprod_{n \geqq 0} \Delta^n \times X_n/\sim$$

with the identifications

(5.6) $$(\varepsilon^i t, x) \sim (t, \varepsilon_i x), \quad t \in \Delta^{n-1},\ x \in X_n,\ i = 0,\ldots,n,\quad n = 1,2,\ldots$$

Remark 1. It is common furthermore to require

(5.7) $$(\eta^i t, x) \sim (t, \eta_i x), \quad t \in \Delta^{n+1},\ x \in X_n,\ i = 0,\ldots,n,\quad n = 0,1,\ldots$$

The resulting space is called the geometric realization and is denoted by $|X|$. One can show that the natural map $\| X \| \to |X|$ is a homotopy equivalence under suitable conditions.

Remark 2. Notice that both $\|\cdot\|$ and $|\cdot|$ are functors.

Example 1. If $X = \{X_q\}$ is a simplicial set then we can consider X as a simplicial space with the discrete topology. The name "geometric realization" for the space $|X|$ originates from this case.

Example 2. Let X be a topological space and let NX be the simplicial space with $NX_q = X$ and all face and degeneracy operators equal to the identity. Then $|NX| = X$ and $\| NX \| = \| N(pt) \| \times X$, where

$$\| N(pt) \| = \Delta^0 \cup \Delta^1 \cup \ldots \cup \Delta^n \cup \ldots$$

with the apropriate identifications.

Example 3. Let G be a Lie group (or more generally any topological group) and consider the following two simplicial spaces $N\bar{G}$ and NG:

$$N\bar{G}(q) = G \times \ldots \times G \quad (q+1\text{-times}),$$
$$NG(q) = G \times \ldots \times G \quad (q\text{-times}).$$

(Here $NG(0)$ consists of one element, namely the empty 0-tuple !).

In $N\bar{G}$ $\varepsilon_i : N\bar{G}(q) \to N\bar{G}(q-1)$ and $\eta_i : N\bar{G}(q) \to N\bar{G}(q+1)$ are given by

$$\varepsilon_i(g_0,\ldots,g_q) = (g_0,\ldots,\hat{g}_i,\ldots,g_q)$$
$$\eta_i(g_0,\ldots,g_q) = (g_0,\ldots,g_{i-1},g_i,g_i,\ldots,g_q), \quad i = 0,\ldots,q.$$

Similarly in NG $\varepsilon_i : NG(q) \to NG(q-1)$ is given by

$$\varepsilon_i(g_1,\ldots,g_q) = \begin{cases} (g_2,\ldots,g_q), & i = 0 \\ (g_1,\ldots,g_i g_{i+1},\ldots,g_q), & i = 1,\ldots,q-1 \\ (g_1,\ldots,g_{q-1}), & i = q \end{cases}$$

and $\eta_i : NG(q) \to NG(q+1)$ by

$$\eta_i(g_1,\ldots,g_q) = (g_1,\ldots,g_{i-1},1,g_i,\ldots,g_q), \quad i = 0,\ldots,q.$$

By definition $EG = \| N\bar{G} \|$ and if we consider the simplicial map $\gamma : N\bar{G} \to NG$ given by

(5.8) $$\gamma(g_0,\ldots,g_q) = (g_0g_1^{-1},\ldots,g_{q-1}g_q^{-1})$$

it is easy to see that there is a commutative diagram

$$\begin{array}{ccc} EG & = & \| N\bar{G} \| \\ {\scriptstyle \gamma_G}\downarrow & & \downarrow{\scriptstyle \|\gamma\|} \\ BG & \longrightarrow & \| NG \| \end{array}$$

such that the bottom horizontal map is a homeomorphism. We will therefore identify BG with $\| NG \|$ and γ_G with $\| \gamma \|$.

The simplicial spaces $N\bar{G}$ and NG above are special cases of the following:

Example 4. Let C be a topological category, i.e. a "small" category such that the set of objects $\mathit{Ob}(C)$ and the set of morphisms $\mathit{Mor}(C)$ are topological spaces and such that

(i) The "source" and "target" maps $\mathit{Mor}(C) \rightrightarrows \mathit{Ob}(C)$ are continuous.

(ii) "Composition": $\mathit{Mor}(C)^{\circ} \to \mathit{Mor}(C)$ is continuous where $\mathit{Mor}(C)^{\circ} \subseteq \mathit{Mor}(C) \times \mathit{Mor}(C)$ consists of the pairs of composable morphisms (i.e. $(f,f') \in \mathit{Mor}(C)^{\circ} \Leftrightarrow$ source (f) = target (f')).

Associated to C there is a simplicial space NC called the nerve of C where $NC(0) = \mathit{Ob}(C)$, $NC(1) = \mathit{Mor}(C)$, $NC(2) = \mathit{Mor}(C)^{\circ}$, and generally

$$NC(n) \subseteq \mathit{Mor}(C) \times \ldots \times \mathit{Mor}(C) \quad (n \text{ times})$$

is the subset of composable strings

$$\cdot \xleftarrow{f_1} \cdot \xleftarrow{f_2} \cdot \longleftarrow \cdot \;\ldots\; \cdot \xleftarrow{f_n} \cdot$$

That is, $(f_1,f_2,\ldots,f_n) \in NC(n)$ iff source (f_i) = target (f_{i+1}), $i = 1,\ldots,n-1$. Here $\varepsilon_i : NC(n) \to NC(n-1)$ is given by

$$\varepsilon_i(f_1,f_2,\ldots,f_n) = \begin{cases} (f_2,\ldots,f_n), & i = 0 \\ (f_1,\ldots,f_i \circ f_{i+1},\ldots,f_n), & 0 < i < n \\ (f_1,\ldots,f_{n-1}), & i = n \end{cases}$$

and $\eta_i : NC(n) \to NC(n+1)$ is given by

$$\eta_i(f_1,\ldots,f_n) = (f_1,\ldots,f_{i-1},id,f_i,\ldots,f_n), \quad i = 0,\ldots,n.$$

Remark 1. Notice that N is a functor from the category of topological categories (where the morphisms are continuous functors) to the category of simplicial spaces.

Remark 2. Observe that a topological group is a topological category with just one object and it follows that NG as defined in Example 3 is exactly the nerve of G as defined in Example 4. Furthermore the simplicial space $N\overline{G}$ defined in Example 3 is exactly the nerve of the category $\overline{G}$ defined as follows: $\mathcal{Ob}(G) = G$ and $\mathcal{Mor}G = G \times G$, source $(g_0,g_1) = g_1$, target $(g_0,g_1) = g_0$ and $(g_0,g_1) \circ (g_1,g_2) = (g_0,g_2)$. Finally $\gamma : N\overline{G} \to NG$ is the nerve of the functor (also called γ) $\gamma : \overline{G} \to G$ given by $\gamma(g_0,g_1) = g_0\, g_1^{-1}$.

Example 5. The following case of Example 4 is useful in the study of G-bundles. Let X be a topological space and $\mathcal{U} = \{U_\alpha\}_{\alpha\in\Sigma}$ an open covering of X. Associated with $\mathcal{U}$ there is a topological category $X_{\mathcal{U}}$ defined as follows: An object is a pair (x,U_α) with $x \in U_\alpha$ and there is a unique morphism $(x,U_{\alpha_0}) \leftarrow (y,U_{\alpha_1})$ iff $x = y \in U_{\alpha_0} \cap U_{\alpha_1}$. That is,

$$\mathcal{Ob}(X_{\mathcal{U}}) = \coprod_{\alpha} U_\alpha, \quad \mathcal{Mor}(X_{\mathcal{U}}) = \coprod_{(\alpha_0,\alpha_1)} U_{\alpha_0} \cap U_{\alpha_1}$$

where the disjoint union is taken over all pairs (α_0, α_1) with $U_{\alpha_0} \cap U_{\alpha_1} \neq \emptyset$. In the simplicial space $NX_{\mathcal{U}}$ the set of n-simplices is

$$NX_{\mathcal{U}}(n) = \coprod_{(\alpha_0, \ldots, \alpha_n)} U_{\alpha_0} \cap \ldots \cap U_{\alpha_n}$$

where again the disjoint union is taken over all (n+1)-tuples $(\alpha_0, \ldots, \alpha_n)$ with $U_{\alpha_0} \cap \ldots \cap U_{\alpha_n} \neq \emptyset$. The face and degeneracy operators are given by natural inclusions. Notice that this simplicial space already appeared in Chapter 1. Notice also that when $\mathcal{U} = \{X\}$ then NX is the simplicial space considered in Example 2.

Now let $\pi : E \to X$ be a topological principal G-bundle (G a Lie group) and let $\mathcal{U} = \{U_\alpha\}$ be an open covering of X with trivializations $\varphi_\alpha : \pi^{-1}(U_\alpha) \to U_\alpha \times G$ and transition functions $g_{\alpha\beta} : U_\alpha \cap U_\beta \to G$. Notice that the cocycle condition (3.2) can be expressed by saying that the transition functions define a continuous functor of topological categories

$$\psi_{\mathcal{U}}(E) : X_{\mathcal{U}} \to G$$

where $\psi_{\mathcal{U}}(E) | U_{\alpha_0} \cap U_{\alpha_1} = g_{\alpha_0\alpha_1}$. Similarly let $\mathcal{V} = \{V_\alpha\}$ be the covering of the total space E by $V_\alpha = \pi^{-1}(U_\alpha)$. Then the trivializations $\{\varphi_\alpha\}$ defines a functor

$$\overline{\psi}_{\mathcal{U}}(E) : E_{\mathcal{V}} \to \overline{G}$$

where

$$\overline{\psi}_{\mathcal{U}}(E) | V_{\alpha_0} \cap V_{\alpha_1} = (\pi_2 \circ \varphi_{\alpha_0}, \pi_2\varphi_{\alpha_1})$$

(here $\pi_2 : V_{\alpha_0} \cap V_{\alpha_1} \times G \to G$ is the projection on the second factor). Finally the projection $\pi : E \to X$ induces a continuous functor $\pi_{\mathcal{U}} : E_{\mathcal{V}} \to X_{\mathcal{U}}$ such that the diagram

$$(5.10)\qquad \begin{array}{ccc} E_{V} & \xrightarrow{\overline{\psi}_{U}} & \overline{G} \\ {\scriptstyle \pi_{U}}\big\downarrow & & \big\downarrow{\scriptstyle \pi_{\gamma}} \\ X_{U} & \xrightarrow{\psi_{U}} & G \end{array}$$

commutes. Also we have the commutative diagram

$$(5.11)\qquad \begin{array}{ccc} E & \longleftarrow & E_{V} \\ {\scriptstyle \pi}\big\downarrow & & \big\downarrow{\scriptstyle \pi_{U}} \\ X & \longleftarrow & X_{U} \end{array}$$

where the horizontal maps are induced by the inclusions. Taking nerves and realizations we get from (5.10) and (5.11) a commutative diagram

$$(5.12)\qquad \begin{array}{ccccc} E = |NE| & \longleftarrow & \| NE_{V} \| & \xrightarrow{\overline{f}_{U}} & \| N\overline{G} \| = EG \\ {\scriptstyle \pi}\big\downarrow & & \big\downarrow{\scriptstyle \| \pi_{U} \|} & & \big\downarrow{\scriptstyle \gamma_{G}} \\ X = |NX| & \xleftarrow{\varepsilon_{U}} & \| NX_{U} \| & \xrightarrow{f_{U}} & \| NG \| = BG \end{array}$$

where $f_{U} = \| \psi_{U}(E) \|$, $\overline{f}_{U} = \| \overline{\psi}_{U}(E) \|$ and $\varepsilon_{U} : \| NX_{U} \| \to X$ is induced by the projection on the second factor in $\coprod_{n} \Delta^{n} \times NX_{U}(n)$. Notice that the upper horizontal maps in (5.12) are equivariant and using Lemma 5.4 it is easily seen that the map $\| \pi_{U} \|$ in the middle is a principal G-bundle. Therefore

$$(5.13)\qquad \varepsilon_{U}^{*}E = f_{U}^{*}E(G).$$

For the proof of Theorem 5.5 we shall study the diagram (5.12) in cohomology. More generally let us study the cohomology of the fat realization of a simplicial space. In the remainder of this chapter we shall use the following notation: For a topological space X, $S_{q}(X) = S_{q}^{top}(X)$ denotes the set of continuous singular q-simplices, and for Λ

a fixed ring $C^q(X) = C^q(X,\Lambda)$ denotes the set of singular cochains with coefficients in Λ and

$$H^q(X) = H^q(X,\Lambda) = H^q(C^*(X,\Lambda)).$$

Now consider a simplicial space $X = \{X_p\}$ and let $C^{p,q}(X)$ denote the double complex

(5.14) $$C^{p,q}(C) = C^q(X_p).$$

Here the vertical differential δ'' is $(-1)^p$ times the coboundary in the complex $C^*(X_p)$ and the horizontal differential δ' is given by

$$\delta' = \sum_{i=0}^{p+1} (-1)^i \varepsilon_i^{\#} : C^q(X_p) \to C^q(X_{p+1}).$$

As in Chapter 1 $C^*(X)$ denotes the total complex of $\{C^{p,q}(X)\}$.

<u>Example 6</u>. If $\mathcal{U} = \{U_\alpha\}_{\alpha\in\Sigma}$ is a covering of a space X then the double complex $C^{p,q}(NX_{\mathcal{U}})$ is exactly the double complex $C_{\mathcal{U}}^{p,q}$ of Chapter 1 (except that in Chapter 1 we considered a C^∞ manifold and C^* denoted C^∞ singular cochains).

Notice that a simplicial map $f = \{f_p\}$ of simplicial spaces $f : X \to X'$ (that is, $f_p : X_p \to X'_p$ is continuous for all p) induces a map of double complexes $f^{\#} : C^{p,q}(X') \to C^{p,q}(X)$. We now have

<u>Proposition 5.15</u>. Let $X = \{X_p\}$ be a simplicial space. Then

$$H^*(\|X\|) \cong H(C^*(X)).$$

Furthermore this isomorphism is natural, i.e. if $f : X \to X'$ is a simplicial map of simplicial spaces then the diagram

$$\begin{array}{ccc} H^*(\|X'\|) & \cong & H(C^*(X')) \\ \| f \|^* \downarrow & & \downarrow f^* \\ H^*(\|X\|) & \cong & H(C^*(X)) \end{array}$$

commutes, where f^* is induced by $f^{\#} : C^*(X') \to C^*(X)$.

Sketch Proof. First assume X is descrete. Then $\|X\|$ is a C.W.-complex with a p-cell for each $x \in X_p$. Therefore the group of cellular p-chains is just $C_p(X)$ and it is straightforward to check that the cellular boundary is given by

$$\partial(\sigma) = \sum_i (-1)^i \varepsilon_i(\sigma), \quad \sigma \in X_p.$$

(For the cellular complex see A. Dold [10, Chapter V, §§ 1 and 6]. It follows that $H^*(\|X\|)$ is naturally isomorphic with the cohomology of the complex $\mathrm{Hom}(C_*(X),\Lambda)$. On the other hand for X discrete $S_q(X_p) = X_p$, $\forall q$, hence

$$C^{p,q}(X) = \mathrm{Hom}(C_p(X),\Lambda), \ \forall q,$$

and the differential $\delta'' : C^{p,q}(X) \to C^{p,q+1}(X)$ is zero for q even and the identity for q odd. Therefore by Corollary 1.20 the natural inclusion

$$C^p(X) = C^{p,0}(X) \subset C^p(X)$$

induces an isomorphism on homology. This proves the proposition in the discrete case.

In particular if Y is a topological space then the natural map $\rho : \|S(Y)\| \to Y$ induces an isomorphism in cohomology (ρ is defined by sending $(t,\sigma) \in \Delta^p \times S_p(Y)$ to $\sigma(t) \in Y$). Notice that by a similar argument ρ induces an isomorphism in homology with Λ coefficients.

Now for a general simplicial space $X = \{X_p\}$ consider the <u>double simplicial set</u> $S(X) = \{S_q(X_p)\}$, that is we have face operators

$$\varepsilon'_i : S_q X_p \to S_q X_{p-1}, \quad \varepsilon''_j : S_q(X_p) \to S_{q-1}(X_p)$$

$i = 0,\ldots,p$, $j = 0,\ldots,q$, such that

$$\varepsilon'_i \circ \varepsilon''_j = \varepsilon''_j \circ \varepsilon'_i$$

and similarly for the degeneracy operators. For this double simplicial set we have the <u>fat</u> realization

$$\| SX \| = \coprod_{p,q\geq 0} \Delta^p \times \Delta^q \times S_q(X_p)/\sim$$

with suitable identifications. Again this is a C.W.-complex and the set of n-cells are in 1-1 correspondence with $\coprod_{p+q=n} S_q(X_p)$. Again one checks that $H^*(\| S(X) \|)$ is isomorphic with $H(C^*(X))$.

On the other hand $\| S(X) \|$ is homoemorphic with the fat realization of the simplicial space $\{\| S(X_p) \|\}$. Now there is a natural simplicial map $\rho = \{\rho_p\}$ where $\rho_p : \| S(X_p) \| \to X_p$ is defined above and, as remarked there, induces an isomorphism in homology. The proposition now follows from the following

<u>Lemma 5.16</u>. Let $f : X \to X'$ be a simplicial map of simplicial spaces such that $f_p : X_p \to X'_p$ induces an isomorphism in homology with coefficients in Λ for all p. Then $\| f \| : \| X \| \to \| X' \|$ also induces an isomorphism in homology as well as in cohomology with coefficients in Λ.

Proof. Let $\| X \|(n) \subseteq \| X \|$ be the image of $\coprod_{k \leq n} \Delta^k \times X_k$. Then $\| X \|(n)$ is a filtration of $\| X \|$ and $\| f \|$ preserves the filtration, that is,

$$\| f \| : \| X \|(n) \to \| X' \|(n).$$

Now it is easy to see that the natural map

$$(\Delta^n \times X_n, \partial\Delta^n \times X_n) \to (\| X \|(n), \| X \|(n-1))$$

induces an isomorphism in homology, hence by assumption

$$\| f \| : (\| X \|(n), \| X \|(n-1)) \to (\| X' \|(n), \| X' \|(n-1))$$

induces an isomorphism in homology. Now iterated use of the five-lemma shows that $\| f \| : \| X \|(n) \to \| X' \|(n)$, $n = 1,2,\ldots$, induces an isomorphism in homology and therefore $\| f \| :$ $\| X \| \to \| X' \|$ also induces an isomorphism in homology. By the Universal coefficient theorem the result now follows, and thus finishes the proof of Proposition 5.15.

Corollary 5.17. Suppose $f_0, f_1 : X \to X'$ are simplicially homotopic simplicial maps of simplicial spaces (i.e., for each p there are continuous maps $h_i : X_p \to X'_{p+1}$, $i = 0,\ldots,p$, satisfying i) - iii) of Exercise 2b) of Chapter 2). Then

$$\| f_0 \|^* = \| f_1 \|^* : H^*(\| X' \|) \to H^*(\| X \|).$$

Proof. In fact consider the induced maps

$$f_0^{\#}, f_1^{\#} : C^{p,q}(X') \to C^{p,q}(X)$$

and let $s_{p+1} : C^{p+1,q}(X') \to C^{p,q}(X)$ be defined by

$$s_{p+1} = \sum_{i=0}^{p} (-1)^i h_i^{\#} .$$

Then

$$s_{p+1} \circ \delta' + \delta' \circ s_p = f_1^{\#} - f_0^{\#}$$

as in Exercise 2 of Chapter 2. Furthermore

$$s_{p+1} \circ \delta'' + \delta'' \circ s_{p+1} = 0$$

since $h_i^{\#}$ are chain maps $C^*(X'_{p+1}) \to C^*(X_p)$. It follows that $f_1^{\#}$ and $f_0^{\#} : C^*(X') \to C^*(X)$ are chain homotopic and hence induce the same map in homology.

<u>Proof of Theorem 5.5</u>. First let c be a characteristic class and let $\pi : E \to X$ be a principal G-bundle. Choose a covering $\mathcal{U}$ of X such that there are trivializations $\varphi_\alpha : \pi^{-1}(U_\alpha) \to U_\alpha \times G$ and consider the diagram (5.12) above. Notice that there is a commutative diagram

$$\begin{array}{ccc} H^*(\| NX_{\mathcal{U}} \|) & \cong & H(C^*(NX_{\mathcal{U}})) \\ {\scriptstyle \varepsilon_{\mathcal{U}}^*} \uparrow & & \| \\ H^*(X) & \xrightarrow{e_C} & H(C_{\mathcal{U}}^*) \end{array}$$

where e_C is the isomorphism of Lemma 1.25, so that $\varepsilon_{\mathcal{U}}^*$ is also an isomorphism.

Now by naturality of c

$$(5.18) \qquad \varepsilon_{\mathcal{U}}^*(c(E)) = f_{\mathcal{U}}^*(c(EG))$$

and since $\varepsilon_{\mathcal{U}}^*$ is an isomorphism $c(E)$ is uniquely determined by $c(EG)$ and equation (5.18).

On the other hand let $c_0 \in H^*(BG)$ and define for a principal G-bundle the class $c(E)$ by

$$(5.19) \qquad \varepsilon_{\mathcal{U}}^*(c(E)) = f_{\mathcal{U}}^*(c_0).$$

we must show that $c(E)$ is well defined:

Now if $\mathcal{U} = \{U_\alpha\}_{\alpha\in\Sigma}$ and $\mathcal{U}' = \{U'_\beta\}_{\beta\in\Sigma'}$ are two coverings of X then $\mathcal{W} = \{U_\alpha \cap U'_\beta\}_{(\alpha,\beta)\in\Sigma\times\Sigma'}$ is also a covering of X and clearly there is a commutative diagram

(5.20)

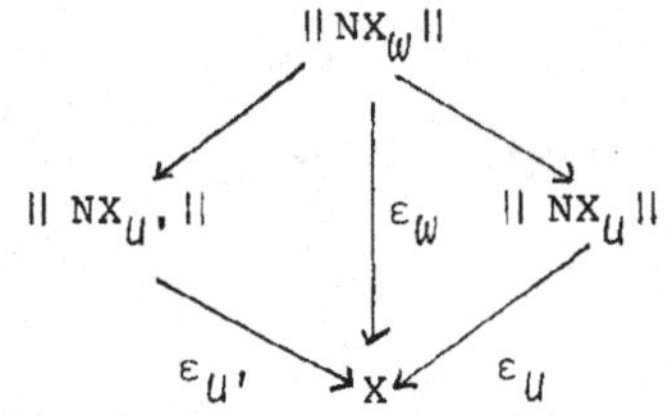

Also let $f_{\mathcal{W}} : \|NX_{\mathcal{W}}\| \to BG$ be the realization of $N\psi_{\mathcal{W}}$ where $\psi_{\mathcal{W}}$ is given by the transition functions corresponding to the trivializations $\varphi_\alpha | U_\alpha \cap U'_\beta$. Then clearly there is a commutative diagram

(5.21)

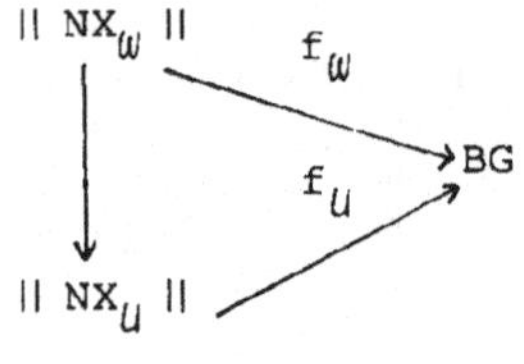

From the diagram (5.20) and (5.21) it follows that it is enough to show that for any covering $\mathcal{U}$ the element $f^*_{\mathcal{U}}(c_0) \in H^*(\|NX_{\mathcal{U}}\|)$ <u>does</u> <u>not</u> depend on the particular choices of trivializations $\{\varphi_\alpha\}$.

So let $\{\varphi_\alpha\}$ and $\{\varphi'_\alpha\}$ be two sets of trivializations associated to $\mathcal{U} = \{U_\alpha\}$ and let $\psi, \psi' : X_{\mathcal{U}} \to G$ be the corresponding continuous functors. We want to show that the associated maps $f_{\mathcal{U}}, f'_{\mathcal{U}} : \|NX_{\mathcal{U}}\| \to BG$ induce the same map in cohomology. Now the family of continuous maps $\lambda_\alpha : U_\alpha \to G$, $\alpha \in \Sigma$, defined by

$$\varphi'_\alpha \circ \varphi_\alpha^{-1}(x,g) = (x,\lambda_\alpha(x)\cdot g), \quad (x,g) \in U_\alpha \times G,$$

satisfy

$$\psi'_{(\alpha_0,\alpha_1)}(x) \cdot \lambda_{\alpha_1}(x) = \lambda_{\alpha_0}(x) \cdot \psi_{(\alpha_0,\alpha_1)}(x), \quad x \in U_{\alpha_0} \cap U_{\alpha_1}.$$

Hence $\lambda = \{\lambda_\alpha\}_{\alpha\in\Sigma}$ is just a <u>continuous natural transformation</u> $\lambda : \psi \to \psi'$ of the functors ψ and ψ'. That $f^*_U = f'^*_U$ therefore follows from Corollary 5.17 and the following general lemma:

<u>Lemma 5.20</u>. Let $\psi,\psi' : C \to D$ be two continuous functors of topological categories C, D. If $\lambda : \psi \to \psi'$ is a continuous natural transformation then $N\psi, N\psi' : NC \to ND$ are simplicially homotopic simplicial maps.

<u>Proof</u>. We shall construct $h_i : NC(p) \to ND(p+1)$, $i = 0,\dots,p$, satisfying i) - iii) of Exercise 2b) in Chapter 2. Now a p-simplex in NC is a string

$$A_0 \xleftarrow{f_1} A_1 \xleftarrow{f_2} A_2 \longleftarrow \cdots\cdots \xleftarrow{f_p} A_p, \quad A_0,\dots,A_p \in \mathcal{Ob}(C),$$
$$f_0,\dots,f_p \in \mathcal{Mor}(C).$$

For $i = 0,\dots,p$, h_i associates to this string the $(p+1)$-simplex in ND given by the string

$$\psi'(A_0) \xleftarrow{\psi'(f_1)} \psi'(A_1) \longleftarrow \cdots \xleftarrow{\psi'(f_i)} \psi'(A_i) \xleftarrow{\lambda_{A_i}} \psi(A_i) \xleftarrow{\psi(f_{i+1})} \cdots$$
$$\cdots \xleftarrow{\psi(f_p)} \psi(A_p).$$

$h_i : NC(p) \to ND(p+1)$ is clearly continuous and it is straightforward to check the identities i) - iii) of Exercise 2b) in Chapter 2. This proves the lemma.

It follows that c(E) defined by (5.19) is well defined and it is easily checked that c(E) satisfies the naturality condition (5.2). This ends the proof of Theorem 5.5.

Note. The original construction of a classifying space is due to J. Milnor [20]. Our exposition follows essentially the one in G. Segal [24].

6. Simplicial manifolds. The Chern-Weil homomorphism for BG

In this chapter H^* again denotes cohomology with real coefficients. We now want to define for a Lie group G the Chern-Weil homomorphism $w : I^*(G) \to H^*(BG)$; but the trouble is that BG is not a manifold. However, $BG = \| N(G) \|$, and NG is a simplicial manifold. That is, $X = \{X_q\}$ a simplicial set is called a simplicial manifold if all X_q are C^∞ manifolds and all face and degeneracy operators are C^∞ maps.

Example 1. Again a simplicial set $X = \{X_q\}$ is a simplicial manifold with all X_q considered as zero dimensional manifolds.

Example 2. Also if M is a C^∞ manifold the simplicial space NM with $NM(q) = M$ and all face and degeneracy operators equal to the identity is again a simplicial manifold.

Example 3. For G a Lie group the simplicial spaces $N\bar{G}$ and NG are also simplicial manifolds and $\gamma : N\bar{G} \to NG$ is a differentiable simplicial map.

Example 4. For M a C^∞ manifold with an open covering $\mathcal{U} = \{U_\alpha\}_{\alpha\in\Sigma}$ the simplicial space $NM_{\mathcal{U}}$ is also a simplicial manifold. Finally, if $\pi : E \to M$ is a differentiable principal G-bundle with differentiable trivializations $\varphi_\alpha : \pi^{-1}U_\alpha \to U_\alpha \times G$ then taking the nerves of the diagrams (5.10) and (5.11) we obtain the corresponding diagrams of simplicial manifolds and differentiable simplicial maps.

Now let us study the cohomological properties of a simplicial manifold, in particular we want a de Rham theorem. Again in this chapter for M a manifold $C^*(M)$ denotes the cochain complex with real coefficients based on C^∞ singular simplices.

Now consider a simplicial manifold $X = \{X_p\}$. As in Chapter 5 we have the double complex $C^{p,q}(X) = C^q(X_p)$. Notice that by Lemma 1.19 and Exercise 4 of Chapter 1 the natural map

$$C^q_{top}(X_p) \to C^q(X_p)$$

induces an isomorphism on homology of the total complexes.

We also have the double complex $A^{p,q}(X) = A^q(X_p)$. Here the vertical differential d'' is $(-1)^p$ times the exterior differential in $A^*(X_p)$ and the horizontal differential $\delta' : A^q(X_p) \to A^q(X_{p+1})$ is defined by

$$\delta' = \sum_{i=0}^{p+1} (-1)^i \varepsilon_i^*.$$

Furthermore we have an integration map

$$I_X = A^{p,q}(X) \to C^{p,q}(X)$$

which is clearly a map of double complexes. By Theorem 1.15 and Lemma 1.19 we easily obtain

<u>Proposition 6.1</u>. Let $X = \{X_p\}$ be a simplicial manifold. Then $I_X : A^{p,q}(X) \to C^{p,q}(X)$ induces a natural isomorphism

$$H(A^*(X)) \cong H(C^*(X)) \cong H^*(\| X \|).$$

Now there is even another double complex associated to a simplicial manifold which generalizes the simplicial de Rham complex of Chapter 2:

Definition 6.2. A simplicial n-form φ on the simplicial manifold $X = \{X_p\}$ is a sequence $\varphi = \{\varphi^{(p)}\}$ of n-forms $\varphi^{(p)}$ on $\Delta^p \times X_p$, such that

$$(6.3) \qquad (\varepsilon^i \times id)^*\varphi^{(p)} = (id \times \varepsilon_i)^*\varphi^{(p-1)} \quad \text{on} \quad \Delta^{p-1} \times X_p, \qquad i = 0,\dots,p, \quad p = 0,1,2,\dots$$

Remark. Notice that $\varphi = \{\varphi^{(p)}\}$ defines an n-form on $\coprod_{p=0}^{\infty} \Delta^p \times X_p$ and that (6.3) is the natural condition for a form on $\|X\|$ in view of the identifications (5.6). In the following the restriction $\varphi^{(p)}$ of φ to $\Delta^p \times X_p$ is also denoted φ. Notice also that for X discrete Definition 6.2 agrees with Definition 2.8.

Let $A^n(X)$ denote the set of simplicial n-forms on X. Again the exterior differential on $\Delta^p \times X_p$ defines a differential $d : A^n(X) \to A^{n+1}(X)$ and also we have the exterior multiplication

$$\wedge : A^n(X) \otimes A^m(X) \to A^{n+m}(X)$$

satisfying the usual identities.

The complex $(A^*(X),d)$ is actually the total complex of a double complex $(A^{k,l}(X),d',d'')$. Here an n-form φ lies in $A^{k,l}(X)$, $k+l = n$ iff $\varphi|\Delta^p \times X_p$ is locally of the form

$$\varphi = \sum a_{i_1\dots i_k,j_1\dots j_l} dt_{i_1} \wedge\dots\wedge dt_{i_k} \wedge dx_{j_1} \wedge\dots\wedge dx_{j_l}$$

where $(t_0,\dots,t_p)$ as usual are the barycentric coordinates in Δ^p and $\{x_j\}$ are local coordinates in X_p. It is easy to see that

$$A^n(X) = \coprod_{k+l=n} A^{k,l}(X)$$

and that

$$d = d' + d''$$

where d' is the exterior derivative with respect to the barycentric coordinates and d'' is $(-1)^k$ times the exterior derivative with respect to the x-variables.

Now restricting a (k,l)-form to $\Delta^k \times X_k$ and integrating over Δ^k yields a map

$$I_\Delta : A^{k,l}(X) \to A^{k,l}(X)$$

which is clearly a map of double complexes. The following theorem is now a strightforward generalization of Theorem 2.16:

<u>Theorem 6.4</u>. For each l the two chain complexes $(A^{*,l}(X), d')$ and $(A^{*,l}(X), \delta')$ are chain equivalent. In fact there are natural maps

$$I_\Delta : A^{k,l}(X) \rightleftarrows A^{k,l}(X) : E$$

and chain homotopies

$$s_k : A^{k,l}(X) \to A^{k-1,l}(X)$$

such that

(6.5) $\quad I_\Delta \circ d' = \delta' \circ I, \qquad I_\Delta \circ d'' = d'' \circ I_\Delta$

(6.6) $\quad d' \circ E = E \circ \delta', \qquad E \circ d'' = d'' \circ E$

(6.7) $\quad I_\Delta \circ E = id$

(6.8) $\quad E \circ I_\Delta - id = s_{k+1} \circ d' + d' \circ s_k, \quad s_k \circ d'' + d'' \circ s_k = 0.$

In particular $I_\Delta : A^{k,l}(X) \to A^{k,l}(X)$ induces a natural isomorphism on the homology of the total complexes

(6.9) $$H(A^*(X)) \cong H(A^*(X)) \cong H^*(\| X \|).$$

Also let us state without proof (see J. L. Dupont [11]) the following generalization of Theorem 2.33:

Theorem 6.10. The isomorphism (6.9) is multiplicative where the product on the left is induced by the $\wedge$-product and where the product on the right is the cup-product.

As an application of Theorem 6.4 let us consider a manifold M with a covering $\mathcal{U} = \{U_\alpha\}_{\alpha\in\Sigma}$ and let $NM_{\mathcal{U}}$ be the simplicial manifold associated to the nerve of the category $M_{\mathcal{U}}$. Notice that the natural map

$$\varepsilon_{\mathcal{U}} : \|NM_{\mathcal{U}}\| \to M$$

is induced by the natural projections

$$\Delta^p \times U_{\alpha_0} \cap \ldots \cap U_{\alpha_p} \to U_{\alpha_0} \cap \ldots \cap U_{\alpha_p} \subseteq M$$

and that these also induce the natural map

$$A^*(M) \to A^*(NM) \to A^*(NM_{\mathcal{U}}) .$$

Corollary 6.11. For $\mathcal{U} = \{U_\alpha\}$ an open covering of M the natural map $A^*(M) \to A^*(NM_{\mathcal{U}})$ induces an isomorphism in homology.

Proof. In fact the composite

$$A^*(M) \longrightarrow A^*(NM_{\mathcal{U}}) \xrightarrow{I_\Delta} A^*(NM_{\mathcal{U}}) = A^*_{\mathcal{U}}$$

is the map e_A of Lemma 1.24.

Now let us turn to Chern-Weil theory for simplicial manifolds. A simplicial G-bundle $\pi : E \to M$ is of course a sequence $\pi_p : E_p \to M_p$ of differentiable G-bundles where $E = \{E_p\}$, $M = \{M_p\}$ are simplicial manifolds, $\pi : E \to M$ is a simplicial differentiable map and also right multiplication by $g \in G$, $R_g : E \to E$, is simplicial. A connection in $\pi : E \to M$ is then a 1-form θ on E (in the sense of Definition 6.2 above) with coefficients in $\mathfrak{g}$ such that θ restricted to $\Delta^p \times E_p$ is a connection in the usual sense in the bundle $\Delta^p \times E_p \to \Delta^p \times M_p$.

Again we have the curvature form Ω defined by 3.14 and for $P \in I^k(G)$ we get $P(\Omega^k) \in A^{2k}(M)$ a closed form representing a class

$$w_E(P) \in H^{2k}(A^*(M)) \cong H^{2k}(\| M \|)$$

such that Theorem 4.3 holds.

In particular let us consider the simplicial G-bundle $\gamma : N\bar{G} \to NG$. There is actually a canonical connection in this bundle constructed as follows:

Let θ_0 be the Maurer-Cartan connection in the bundle $G \to pt$. Also let $q_i : \Delta^p \times N\bar{G}(p) \to G$ be the projection onto the i-th factor in G^{p+1}, $i = 0,\dots,p$, and let $\theta_i = q_i^*\theta_0$. Then θ is simply given over $\Delta^p \times N\bar{G}(p)$ by

$$(6.12) \qquad \theta = t_0\theta_0 + \dots + t_p\theta_p$$

where as usual $(t_0,\dots,t_p)$ are the barycentric coordinates in Δ^p. By Proposition 3.10, $\theta | \Delta^p \times N\bar{G}(p)$ is clearly a connection in the usual sense and it is also obvious from (6.12) that θ satisfies (6.3). We now summarize:

Theorem 6.13. a) There is a canonical homomorphism

$$w : I^*(G) \to H^*(BG)$$

such that for $P \in I^k(G)$, $w(P)$ is represented in $A^{2k}(NG)$ by

$P(\Omega^k)$ where Ω is the curvature form of the connection θ defined by (6.12).

b) Let for $P \in I^k(G)$, $w(P)(\cdot)$ be the corresponding characteristic class. Then if $\pi : E \to M$ is an ordinary differentiable G-bundle we have

$$w(P)(E) = w_E(P)$$

where $w_E : I^*(G) \to H^*(M)$ is the usual Chern-Weil homomorphism.

c) $w : I^*(G) \to H^*(BG)$ is an algebra homomorphism.

d) Let $\alpha : H \to G$ be a Lie group homomorphism and let $\alpha^* : I^*(G) \to I^*(H)$ be the induced map. Then the diagram

$$\begin{array}{ccc} I^*(G) & \xrightarrow{\alpha^*} & I^*(H) \\ {\scriptstyle w}\downarrow & & \downarrow{\scriptstyle w} \\ H^*(BG) & \xrightarrow{B\alpha^*} & H^*(BH) \end{array}$$

commutes.

Proof. a) is a definition.

b) Choose an open covering $\mathcal{U} = \{U_\alpha\}$ of M and trivializations of E so that we have a commutative diagram of differentiable simplicial bundles:

$$\begin{array}{ccccc} NE & \longleftarrow & NE_{\mathcal{U}} & \longrightarrow & N\bar{G} \\ \downarrow{\scriptstyle N\pi} & & \downarrow{\scriptstyle N\pi_{\mathcal{U}}} & & \downarrow{\scriptstyle \gamma} \\ NM & \longleftarrow & NM_{\mathcal{U}} & \longrightarrow & NG \end{array} .$$

By the proof of Theorem 5.5 the pull back of $w(P)(E)$ to $\| NM_{\mathcal{U}} \|$ is given by $f_{\mathcal{U}}^*(w(P))$ which clearly is represented in $H(A^*(NM_{\mathcal{U}})$ by the Chern-Weil image of P for the simplicial

G-bundle $NE_V \to NM_U$ with connection θ' induced from the connection θ defined by (6.12). On the other hand a connection in $E \to M$ induces another connection θ'' in $NE_V \to NM_U$ and the pull-back of $w_E(P)$ in $H(A^*(NM_U))$ is clearly represented by the Chern-Weil image of P using the connection θ''. However, by the argument of Theorem 4.3, b) the Chern-Weil image is independent of the choice of connection, which proves that

$$\varepsilon_U^*(w(P)(E)) = \varepsilon_U^*(w_E(P)),$$

where $\varepsilon_U : \| NM_U \| \to M$ is the natural map considered above. Since ε_U induces an isomorphism in cohomology this ends the proof of b).

c) follows again from the simplicial analogue of Theorem 4.3 c) and Theorem 6.10.

d) is straightforward and the proof is left to the reader.

Note. Notice that by a), $w(P)$ is also represented in the total complexes $A^*(NG)$ and $C^*(NG)$ by canonically defined elements. The construction of $w(P)$ in $A^*(NG)$ is due to H. Shulman [26] generalizing a construction by R. Bott (see [2], [4], and [5]). The exposition in terms of simplicial manifolds follows J. L. Dupont [11].

7. Characteristic classes for some classical groups

We shall now study the properties of the characteristic classes defined in the examples of Chapter 4.

Chern classes.

For $G = Gl(n,\mathbb{C})$ we considered in Chapter 4 Example 4 the complex valued invariant polynomials C_k, $k = 0,1,\ldots,n$ $(C_0 = 1)$, defined by (4.13). For a differentiable $Gl(n,\mathbb{C})$-bundle $\pi : E \to M$ we thus define characteristic classes called the <u>Chern classes</u>

(7.1) $$c_k(E) = w_E(C_k) \in H^{2k}(M,\mathbb{C}), \quad k = 0,1,\ldots,n,$$

represented by the complex valued 2k-forms $C_k(\Omega^k)$, where Ω is the curvature form of a connection in $\pi : E \to M$. Notice that since every complex vector bundle has a Hermitian metric, i.e. a reduction to $U(n)$, $c_k(E)$ actually lies in the image of the inclusion $H^{2k}(M,\mathbb{R}) \subseteq H^{2k}(M,\mathbb{C})$ (cf. Exercise 4 of Chapter 4).

By Theorem 6.13 we can extend the definition of the Chern classes to any <u>topological</u> $Gl(n,\mathbb{C})$-bundle by first defining

$$c_k = w(C_k) \in H^{2k}(B\,Gl(n,\mathbb{C}),\mathbb{C})$$

and then use Theorem 5.5. Again c_k is a <u>real</u> class. In fact since C_k restricted to $\mathfrak{u}(n)$ is a real polynomial it follows from Theorem 6.13 d) that the restriction of c_k to $BU(n)$ is a real class (represented by a real valued form), and since the natural inclusion $j : U(n) \subset Gl(n,\mathbb{C})$ is a homotopy equivalence it follows that

$$Bj : BU(n) \to B\,Gl(n,\mathbb{C})$$

induces an isomorphism in cohomology. In general we have

Proposition 7.2. Let $\alpha : H \to G$ be a homomorphism of two Lie groups which induces an isomorphism in homology (coefficients in a P.I.D. Λ). Then also $B\alpha : BH \to BG$ induces an isomorphism in homology as well as in cohomology (with coefficients Λ).

Proof. By Künneth's formula $N\alpha(p) : NH(p) \to NG(p)$ induces an isomorphism in homology for each p. The proposition therefore follows by Lemma 5.16.

Before continuing the study of the Chern classes we make a few definitions:

Suppose we consider a topological space X with a principal $Gl(n,\mathbb{C})$-bundle $\xi : E \to X$ and a $Gl(m,\mathbb{C})$-bundle $\zeta : F \to X$. Then the Whitney sum $(\xi \oplus \zeta) : E \oplus F \to X$ is most easily described in terms of transition functions as follows:

First let

$$\oplus : Gl(n,\mathbb{C}) \times Gl(m,\mathbb{C}) \to Gl(n+m,\mathbb{C})$$

be the homomorphism taking a pair of matrices (A,B) to the matrix

$$A \oplus B = \begin{pmatrix} A & 0 \\ 0 & B \end{pmatrix} .$$

Now choose a covering $\mathcal{U} = \{U_\alpha\}_{\alpha \in \Sigma}$ of X such that both E and F have trivializations over U_α, $\alpha \in \Sigma$, and let $\{g_{\alpha\beta}\}$ and $\{h_{\alpha\beta}\}$ be the corresponding transition functions for E and F respectively. Then $\xi \oplus \zeta : E \oplus F \to X$ is the bundle with transition functions $\{g_{\alpha\beta} \oplus h_{\alpha\beta}\}$. Notice that if E and F are differentiable then also $E \oplus F$ is.

Notice that $Gl(1,\mathbb{C}) = \mathbb{C}^* = \mathbb{C} \setminus \{0\}$, the multiplicative group of non-zero complex numbers. $Gl(1,\mathbb{C})$-bundles are in 1-1 correspondence with 1-dimensional complex vector bundles (also called <u>complex line bundles</u>). An important example is the <u>canonical line bundle</u> on the <u>complex projective space</u> $\mathbb{C}P^n$. Here $\mathbb{C}P^n$ is defined as the quotient space of $\mathbb{C}^{n+1} \setminus \{0\}$ under the action of $\mathbb{C}^*$ given by

$$(z_0, z_1, \ldots, z_n) \cdot \lambda = (z_0 \cdot \lambda, \ldots, z_n \cdot \lambda),$$
$$z_0, \ldots, z_n \in \mathbb{C},\ \lambda \in \mathbb{C}^*.$$

It is easy to see that the natural projection

$$\eta_n : \mathbb{C}^{n+1} \setminus \{0\} \to \mathbb{C}P^n$$

is a principal $\mathbb{C}^*$-bundle. The associated complex line bundle is by definition the <u>canonical line bundle</u>. The total space is denoted H_n^* (H for H. Hopf).

We can now prove

<u>Theorem 7.3</u>. For a $Gl(n,\mathbb{C})$-bundle $\xi : E \to X$ let the total <u>Chern class</u> be the sum

$$c(E) = c_0(E) + c_1(E) + \ldots + c_n(E) \in H^*(X,\mathbb{C}).$$

Then

a) $c_i(E) \in H^{2i}(X,\mathbb{C})$, $i = 0,1,\ldots$

$c_0(E) = 1$ and $c_i(E) = 0$ for $i > n$.

b) (Naturality). If $f : Y \to X$ is continuous and $\xi : E \to X$ a $Gl(n,\mathbb{C})$-bundle then

$$c(f^*E) = f^*(c(E)). \tag{7.4}$$

c) (Whitney duality formula). If $\xi : E \to X$ is a $Gl(n,\mathbb{C})$-bundle and $\zeta : F \to X$ a $Gl(m,\mathbb{C})$-bundle then

$$c(E \oplus F) = c(E) \cdot c(F) \tag{7.5}$$

or equivalently

$$c_k(E \oplus F) = \sum_{i+j=k} c_i(E) \smile c_j(F), \quad k = 0,1,\dots,n+m. \tag{7.5'}$$

d) (Normalization). Let $\eta_n : H_n^* \to \mathbb{C}P^n$ be the canonical line bundle. Then

$$c(H_n^*) = 1 - h_n \tag{7.6}$$

where $h_n \in H^2(\mathbb{C}P^n, \mathbb{Z})$ is the canonical generator.

Proof. a) is trivial by definition.

b) follows from Theorem 5.5.

c) Let us write $G_n = Gl(n,\mathbb{C})$ for short. The map $\oplus : G_n \times G_m \to G_{n+m}$ is clearly a homomorphism and the Whitney sum $E \oplus F$ by definition has a reduction to $G_n \times G_m$. $\oplus$ together with the projections

$$p_1 : G_n \times G_m \to G_n, \quad p_2 : G_n \times G_m \to G_m$$

induce the maps in the diagram

$$\begin{array}{ccc} B(G_n \times G_m) & \xrightarrow{B\oplus} & BG_{n+m} \\ \downarrow{\scriptstyle Bp_1 \times Bp_2} & & \\ BG_n \times BG_m & & \end{array} \tag{7.7}$$

and (7.5)' will clearly follow if we can prove the formula

$$(B\oplus)^* c_k = \sum_{i+j=k} (Bp_1)^* c_i \smile (Bp_2)^* c_j, \quad k = 0,1,\dots,n+m. \tag{7.8}$$

We shall prove this by proving the corresponding formula on the level of differential forms in the diagram

$$(7.9)\qquad \begin{array}{ccc} & N(G_n \times G_m) & \xrightarrow{N\oplus} NG_{n+m} \\ & {}^{Np_1}\swarrow \quad \searrow^{Np_2} & \\ NG_n & & NG_m \end{array}$$

For this we first need some notation. Let M_n denote the Lie algebra of G_n (i.e. M_n is the Lie algebra of $n \times n$ matrices) and let $\theta_{(n)}$ be the canonical connection in $N\bar{G}_n$ defined by (6.12) with $\Omega_{(n)}$ the corresponding curvature form. Also let

$$i_1 : M_n \to M_{n+m}, \quad i_2 : M_m \to M_{n+m}$$

be the inclusions given by

$$i_1(A) = \begin{pmatrix} A & 0 \\ 0 & 0 \end{pmatrix}, \quad i_2(B) = \begin{pmatrix} 0 & 0 \\ 0 & B \end{pmatrix}.$$

Then it is easy to see that

$$(7.10)\qquad (N\bar{\oplus})^*\theta_{(n+m)} = (N\bar{p}_1)^*(i_1 \circ \theta_{(n)}) + (N\bar{p}_2)^*(i_2 \circ \theta_{(m)})$$

and since the Lie bracket of the two forms on the right of (7.10) is zero it follows that

$$(7.11)\qquad (N\bar{\oplus})^*\Omega_{(n+m)} = (N\bar{p}_1)^*(i_1 \circ \Omega_{(n)}) + (N\bar{p}_2)^*(i_2 \circ \Omega_{(m)}).$$

Now for $A \in M_n$ and $B \in M_m$

$$\det(\lambda 1 - \frac{1}{2\pi i}(i_1(A) + i_2(B)))$$

$$= \det\begin{pmatrix} \lambda 1 - \frac{1}{2\pi i} A & 0 \\ 0 & \lambda 1 - \frac{1}{2\pi i} B \end{pmatrix}$$

$$= \det(\lambda 1 - \frac{1}{2\pi i} A)\det(\lambda 1 - \frac{1}{2\pi i} B)$$

from which we conclude

$$(7.12)\qquad C_k(i_1A + i_2B,\dots,i_1A + i_2B) = \sum_{i+j=k} C_i(A,\dots,A)\cdot C_j(B,\dots,B).$$

Therefore by (7.11) we have

$$(7.13)\qquad (N\Theta)^*C_k(\Omega^k) = \sum_{i+j=k} Np_1^*C_i(\Omega^i) \wedge Np_2^*C_j(\Omega^j),\ k = 0,1,\dots,n+m,$$

which clearly implies (7.8) and ends the proof of c).

d) The restriction map $H^2(\mathbb{C}P^n,\mathbb{Z}) \to H^2(\mathbb{C}P^1,\mathbb{Z})$ is an isomorphism and h_n maps to h_1, which by definition is the class such that

$$\langle h_1,[\mathbb{C}P^1]\rangle = 1$$

where $\mathbb{C}P^1$ is given the canonical orientation determined by the 2-form $dx \wedge dy$ where $z = x + iy = z_1/z_0$ is the complex coordinate in the Riemann sphere $\mathbb{C}P^1$ with homogeneous coordinates (z_0,z_1).

By naturality it is clearly enough to prove (7.6) for $n = 1$, so we consider the principal $\mathbb{C}^*$-bundle

$$\eta_1 : \mathbb{C}^2 \setminus \{0\} \to \mathbb{C}P^1$$

which is clearly a differentiable bundle. Let (z_0,z_1) be the coordinates in $\mathbb{C}^2 \setminus \{0\}$ and consider the complex valued 1-form

$$(7.14)\qquad \theta = (\bar{z}_0 dz_0 + \bar{z}_1 dz_1)/(|z_0|^2 + |z_1|^2)$$

where the bar denotes complex conjugation and $|z|^2 = z\bar{z}$. Then it is easily checked that θ is a connection and since $\mathbb{C}^*$ is abelian the curvature form is given by

$$(7.15)\qquad \Omega = d\theta.$$

Now let $U = \mathbb{C}P^1\setminus\{(0,1)\} = \{(z_0,z_1) \mid z_0 \neq 0\}$ and use the local coordinate $z = z_1/z_0$. Then $z_1 = z_0 z$ and $dz_1 = z\,dz_0 + z_0\,dz$.

Hence

$$\theta = [\bar{z}_0 dz_0 + \bar{z}_0\bar{z}(zdz_0 + z_0 dz)]/|z_0|^2(1 + |z|^2)$$

$$= (\frac{dz_0}{z_0} + |z|^2 \frac{dz_0}{z_0} + \bar{z}dz)/(1 + |z|^2) = \frac{dz_0}{z_0} + \frac{\bar{z}}{1+|z|^2} dz.$$

Therefore

$$\Omega = d\theta = d(\frac{\bar{z}}{1+z\bar{z}} dz) = \frac{d\bar{z} \wedge dz}{(1+|z|^2)^2} .$$

It follows that in U $C_1(\Omega)$ is given by

$$C_1(\Omega) = -\frac{1}{2\pi i} \frac{d\bar{z} \wedge dz}{(1+|z|^2)^2}.$$

Therefore (cf. Exercise 2 b) below)

$$<c_1(H_1^*),[\mathbb{C}P^1]> = \int_{\mathbb{C}P^1} C_1(\Omega) = -\frac{1}{2\pi i} \int_{\mathbb{C}} \frac{d\bar{z} \wedge dz}{(1+|z|^2)^2} .$$

Now put $z = r\, e^{2\pi i t}$. Then $d\bar{z} \wedge dz = 4\pi i r dr \wedge dt$; Hence

$$<c_1(H_1^*),[\mathbb{C}P^1]> = -2 \int_0^1\int_0^\infty \frac{r}{(1+r^2)^2} dr\, dt$$

$$= -\int_0^\infty \frac{dr}{(1+r)^2} = -1.$$

This proves (7.6) and ends the proof of the theorem.

Remark. In the next chapter we shall see that Theorem 7.3 characterizes the Chern classes uniquely. By topological methods one can show that there exist classes $c_k \in H^{2k}(BGl(n,\mathbb{C}),\mathbb{Z})$ such that the corresponding characteristic classes satisfy Theorem 7.2. It follows that these map to our Chern classes under the natural map induced by $\mathbb{Z} \to \mathbb{C}$.

Pontrjagin classes.

For $G = Gl(n,\mathbb{R})$ we considered in Chapter 4 Example 1 the realvalued invariant polynomials $P_{k/2}$, $k = 0,\ldots,n$, defined by (4.11). For $\pi : E \to M$ a differentiable $Gl(n,\mathbb{R})$-bundle we defined the Pontrjagin classes

(7.16) $p_{k/2}(E) = w_E(P_{k/2}) \in H^{2k}(M,\mathbb{R})$, $k = 0,1,\ldots,n$,

represented by the 2k-forms $P_{k/2}(\Omega^k)$, where Ω is the curvature form of a connection. As noticed in Chapter 4 Example 2, $p_{k/2}(E) = 0$ for k odd. Again we extend the definition to all topological $Gl(n,\mathbb{R})$-bundles by defining

$$p_{k/2} = w(P_{k/2}) \in H^{2k}(B\,Gl(n,\mathbb{R}),\mathbb{R}), \quad k = 0,1,\ldots,n,$$

and using Theorem 5.5. This time the inclusion $j : O(n) \to Gl(n,\mathbb{R})$ is a homotopy equivalence hence by Proposition 7.2 induces an isomorphism

$$(Bj)^* : H^*(B\,Gl(n,\mathbb{R}),\mathbb{R}) \xrightarrow{\cong} H^*(BO(n),\mathbb{R}),$$

and since for k odd $P_{k/2}$ restricted to the Lie algebra $\mathfrak{o}(n)$ is zero it follows that $p_{k/2} = 0$ for k odd.

The proof of the following theorem is left to the reader.

Theorem 7.17. For a $Gl(n,\mathbb{R})$-bundle $\xi : E \to X$ let the total Pontrjagin class be the sum

$$p(E) = p_0(E) + p_1(E) + \ldots + p_{[n/2]}(E) \in H^*(X,\mathbb{R}).$$

Then

a) $p_i(E) \in H^{4i}(X,\mathbb{R})$, $i = 0,1,\ldots$

$p_0(E) = 1$ and $p_i(E) = 0$ for $i > n/2$.

b) Let $\xi_{\mathbb{C}} : E_{\mathbb{C}} \to X$ be the complexification of $\xi : E \to X$, that is, the extension to $Gl(n,\mathbb{C})$. Then

(7.18) $p_i(E) = (-1)^i\, c_{2i}(E_{\mathbb{C}})$, $i = 0,1,\ldots$

c) (Naturality). If $f : Y \to X$ is continuous and

$\xi : E \to X$ is a $Gl(n,\mathbb{R})$-bundle then

(7.19) $$p(f^*E) = f^*p(E).$$

d) (Whitney duality formula). If $\xi : E \to X$ is a $Gl(n,\mathbb{R})$-bundle and $\zeta : F \to X$ a $Gl(m,\mathbb{R})$-bundle then

(7.20) $$p(E \oplus F) = p(E) \cdot p(F),$$

or equivalently

(7.20)' $$p_k(E \oplus F) = \sum_{i+j=k} p_i(E) \smile p_j(F), \quad k = 0,1,2,\ldots,[(n+m)/2].$$

<u>The Euler class</u>.

Finally consider $G = SO(2m)$. In Chapter 4 Example 3 we defined the invariant polynomial Pf by the Equation (4.12). For a differentiable $SO(2m)$ bundle $\pi : E \to M$ we define the <u>Euler class</u>

(7.21) $$e(E) = w_E(Pf) \in H^{2m}(M,\mathbb{R}).$$

Again we extend the definition to topological bundles by putting

$$e = w(Pf) \in H^{2m}(BSO(2m),\mathbb{R})$$

and using Theorem 5.5. We then have

<u>Theorem 7.22</u>. For $\xi : E \to X$ a $SO(2m)$-bundle $e(E) \in H^{2m}(X,\mathbb{R})$ satisfies

a) (Naturality). For $f : Y \to X$ continuous and $\xi : E \to X$ a $SO(2m)$-bundle

(7.23) $$e(f^*E) = f^*e(E).$$

b) (Whitney duality formula). For $\xi : E \to X$ a $SO(2m)$-bundle and $\zeta : F \to X$ a $SO(2l)$-bundle

(7.24) $$e(E \oplus F) = e(E) \cup e(F).$$

c) For $\xi : E \to X$ a $U(m)$-bundle let $\xi_{\mathbb{R}} : E_{\mathbb{R}} \to X$ be the <u>realification</u>, i.e. the extension to $SO(2m)$ (where the inclusion $U(m) \subset SO(2m)$ is defined by identifying $\mathbb{C}^m = \mathbb{R} \oplus i\mathbb{R} \oplus \mathbb{R} \oplus i\mathbb{R} \oplus \ldots \oplus i\mathbb{R} = \mathbb{R}^{2m}$). Then

(7.25) $$e(E_{\mathbb{R}}) = c_m(E).$$

d) For $\xi : E \to X$ an $SO(2m)$-bundle

(7.26) $$e(E)^2 = p_m(E).$$

<u>Proof</u>. a) is trivial by Theorem 5.5.

b) First observe that for $A \in \mathscr{A}(m)$ and $B \in \mathscr{A}(l)$ (that is, A and B are skew-symmetric matrices)

(7.27) $$\mathrm{Pf}\left[\begin{pmatrix} A & 0 \\ 0 & B \end{pmatrix}, \ldots, \begin{pmatrix} A & 0 \\ 0 & B \end{pmatrix}\right] = \mathrm{Pf}(A, \ldots, A) \cdot \mathrm{Pf}(B, \ldots, B).$$

To see this notice that since Pf is invariant it is enough to consider A and B of the form

$$A = \begin{pmatrix} 0 & a_1 & & & \\ -a_1 & 0 & & & 0 \\ & & \ddots & & \\ & & & 0 & a_m \\ & 0 & & -a_m & 0 \end{pmatrix} \qquad B = \begin{pmatrix} 0 & b_1 & & & \\ -b_1 & 0 & & & 0 \\ & & \ddots & & \\ & & & 0 & b_l \\ & 0 & & -b_l & 0 \end{pmatrix}$$

Then clearly

$$\mathrm{Pf}(A, \ldots, A) = \frac{a_1 \ldots a_m}{(2\pi)^m}, \qquad \mathrm{Pf}(B, \ldots, B) = \frac{b_1 \ldots b_l}{(2\pi)^l}$$

and

$$\mathrm{Pf}\left[\begin{pmatrix} A & 0 \\ 0 & B \end{pmatrix}, \ldots, \begin{pmatrix} A & 0 \\ 0 & B \end{pmatrix}\right] = \frac{a_1 \ldots a_m b_1 \ldots b_l}{(2\pi)^{m+l}}$$

so that (7.27) is obvious in this case. Now (7.24) follows from (7.27) exactly as in the proof of Theorem 7.3 c).

c) The inclusion $r : U(m) \subset SO(2m)$ correspond to the map of Lie algebras $r_* : \mathfrak{u}(m) \to \mathfrak{o}(2m)$ which sends the skew-Hermitian $m \times m$-matrix $X = \{a_{st} + ib_{st}\}$ into the $2m \times 2m$-matrix

$$r_*(X) = \begin{pmatrix} a_{11} & -b_{11} & a_{12} & -b_{12} & \cdots\cdots & a_{1m} & -b_{1m} \\ b_{11} & a_{11} & b_{12} & a_{12} & & b_{1m} & a_{1m} \\ & & & & & & \\ a_{m1} & -b_{m1} & & \cdots\cdots\cdots\cdots & & a_{mm} & -b_{mm} \\ b_{m1} & a_{m1} & & & & b_{mm} & a_{mm} \end{pmatrix}$$

which is clearly skew-symmetric. Now (7.25) follows from Theorem 6.13 d) and the following identity of polynomials:

(7.28) $$Pf(r_*(X),\ldots,r_*(X)) = C_m(X,\ldots,X), \quad X \in \mathfrak{u}(m).$$

Since both sides are invariant polynomials on $\mathfrak{u}(m)$ we can again assume that X is diagonalized, that is,

$$X = \begin{pmatrix} ib_1 & & 0 \\ & \ddots & \\ 0 & & ib_m \end{pmatrix}, \quad b_1 \ldots b_m \in \mathbb{R}.$$

Then

$$Pf(r_*(X),\ldots,r_*(X)) = (-1)^m \frac{b_1 \ldots b_m}{(2\pi)^m}$$

whereas

$$C_m(X,\ldots,X) = \det(-\frac{1}{2\pi i} X) = (-1)^m \frac{b_1 \ldots b_m}{(2\pi)^m}$$

which proves (7.28) and hence (7.25).

d) clearly follows from the identity

(7.29) $\quad \mathrm{Pf}(A,\ldots,A)^2 = P_m(A,\ldots,A) = \det(-\frac{1}{2\pi}A), \quad A \in \mathfrak{o}(m),$

which is proved in the same way as (7.27).

Remark. Usually in Algebraic Topology the Euler class is defined differently (see Exercise 1 below). But as we shall see in the next chapter it is uniquely determined by the properties of Theorem 7.22.

Exercise 1. This exercise deals with the algebraic topological definition of the Euler class. In the following H^* denotes cohomology with coefficients in $\mathbb{Z}$. Let $Gl(n,\mathbb{R})^+ \subseteq Gl(n,\mathbb{R})$ be the subgroup of matrices with positive determinant

a) Show that topological $Gl(n,\mathbb{R})^+$-bundles on a topological space X correspond bijectively to oriented vector bundles of dimension n, i.e. n-dimensional vector bundles $\pi : E \to X$ with a preferred orientation of every fibre $E_x = \pi^{-1}(x)$, $x \in X$, such that for every point of X there is a neighbourhood U and a trivialization $\varphi : \pi^{-1}(U) \to U \times \mathbb{R}^n$ which is orientation preserving on every fibre ($\mathbb{R}^n$ is given the canonical orientation).

Now let $E_0 = E \setminus (X \times 0)$, where $X \times 0$ denotes the zero section of E. Recall (see e.g. J. Milnor and J. Stasheff [19, Theorem 9.1]) that there is a unique class $U \in H^n(E,E_0)$ (the Thom class of E) such that for every $x \in X$ and for $i_x : \mathbb{R}^n \to E$ an orientation preserving isomorphism onto the fibre E_x, the class $i_x^*U \in H^n(\mathbb{R}^n,\mathbb{R}^n \setminus \{0\})$ is the canonical generator.

Now let $Y \subseteq X$ and suppose $s : X \to E$ is a section with $s(x) \neq 0$ for all $x \in Y$. Define the relative Euler class

(7.30) $\quad e(E,s) = s^*U \in H^n(X,Y).$

b) Show that $e(E,s)$ does not depend on $s|X-Y$. In particular for $Y = \emptyset$

$$e(E) = e(E,s) \in H^n(X) \tag{7.31}$$

is independent of s (so we can choose s = zero section). Furthermore, show that $e(E)$ only depends on the isomorphism class of E as oriented vector bundle.

c) Observe that for $X = \mathbb{R}^n$, $Y = \mathbb{R}^n \setminus \{0\}$, $E = X \times \mathbb{R}^n$

$$e(E,s) \in H^n(\mathbb{R}^n, \mathbb{R}^n \setminus \{0\}) \cong \mathbb{Z}$$

is just the canonical generator times the degree of $\tilde{s} : \mathbb{R}^n \setminus \{0\} \to \mathbb{R}^n - \{0\}$, where $s(x) = (x, \tilde{s}(x))$, $x \in \mathbb{R}^n$.

d) Let $X = M$ be a compact oriented n-dimensional differentiabel manifold and let $\pi : E \to M$ be an n-dimensional oriented vector bundle on M. Suppose $s : M \to E$ is a section such that s vanishes only at a finite set of points $A_1, \ldots, A_N$. Now choose disjoint neighbourhoods U_i of A_i together with orientation preserving diffeomorphisms $\varphi_i : U_i \to \mathbb{R}^n$ taking A_i to 0 and together with orientation preserving trivializations $\psi_i : \pi^{-1}(U_i) \to U_i \times \mathbb{R}^n$. Clearly $s_i = \psi_i \circ s \circ \varphi_i^{-1}$ defines a section as in c) and we define the integer (the <u>local index</u>)

$$\mathrm{Index}_{A_i}(s) = \deg(\tilde{s}_i). \tag{7.32}$$

Show that $\mathrm{Index}_{A_i}(s)$ is independent of the choices of U_i, φ_i, ψ_i, and show the following formula of H. Hopf:

$$\sum_{i=1}^{N} \mathrm{Index}_{A_i}(s) = \langle e(E), [M] \rangle. \tag{7.33}$$

In particular the left hand side of (7.33) is independent of s.

For the tangent bundle $\tau_M : TM \to M$ one can use (7.33) to show that

$$(7.34) \qquad \langle e(TM),[M]\rangle = \chi(M) = \sum_{i=0}^{n} (-1)^i \dim_{\mathbb{Q}} H_i(M,\mathbb{Q}),$$

the Euler-Poincaré characteristic of M. In fact e.g. the gradient vector field of a Morse function is easily seen to have the sum of local indices equal to $\chi(M)$ (see J. Milnor [21, Theorem 5.2]).

e) Show that $e(E) \in H^n(X)$, defined for $\pi : E \to X$ an oriented vector bundle of dimension n, has the following properties:

i) (Naturality). For $f : Y \to X$ continuous and $\pi : E \to X$ an oriented vector bundle

$$(7.35) \qquad e(f^*E) = f^*e(E),$$

hence e defines a characteristic class with $\mathbb{Z}$-coefficients for principal $Gl(n,\mathbb{R})^+$-bundles.

ii) (Whitney duality formula). For $\pi : E \to X$ an oriented n-dimensional vector bundle and $\xi : F \to X$ an oriented m-dimensional vector bundle

$$(7.36) \qquad e(\xi \oplus \xi) = e(\xi) \cup e(\xi).$$

iii) For $\pi : E \to X$ an oriented vector bundle let $\pi : E^- \to X$ be the vector bundle with the opposite orientation. Then

$$(7.37) \qquad e(E^-) = -e(E).$$

iv) For $\pi : E \to X$ n-dimensional with n <u>odd</u>

$$e(E) \in H^n(X) \quad \text{has order } 2.$$

(Hint: Notice that the antipodal map on each fibre defines an isomorphism of E and E^-).

v) For $\eta_n : H_n^* \to \mathbb{C}P^n$ the canonical complex line bundle considered as a plane bundle with the induced orientation (coming from the usual identification $\mathbb{C} = \mathbb{R} \oplus i\mathbb{R} = \mathbb{R}^2$)

$$(7.38) \qquad \mathcal{e}(H_n^*) = -h_n$$

where $h_n \in H^2(\mathbb{C}P^n)$ is the canonical generator.
(Hint: Use (7.33) for the bundle $\eta_1 : H_1^* \to \mathbb{C}P^1$).

Remark. In the next chapter we shall show that i) - v) determines the image of $\mathcal{e}(E)$ in real cohomology. Hence, by Theorem 7.22,

$$(7.39) \qquad \mathcal{e}(E) = e(E) \in H^{2m}(X,\mathbb{R})$$

for any $SO(2m)$-bundle $\pi : E \to X$.

Exercise 2. Let M be an n-dimensional compact oriented differentiable manifold. The fundamental class $[M] \in H_n(M,\mathbb{Z})$ is by definition the unique class such that for any orientation preserving diffeomorphism $\varphi : \mathbb{R}^n \to U \subseteqq M$ and for $x = \varphi(0)$

$$\varphi_* : H_n(\mathbb{R}^n,\mathbb{R}^n \setminus \{0\},\mathbb{Z}) \to H_n(U,U\setminus\{x\};\mathbb{Z}) \cong H_n(M,M\setminus\{x\};\mathbb{Z})$$

takes the canonical generator to the image of $[M]$ under the natural map $H_n(M;\mathbb{Z}) \to H_n(M,M\setminus\{x\};\mathbb{Z})$. Choose a C^∞ singular n-chain representing $[M]$ and denote it also by $[M]$.

a) As usual let $\Delta^n \subseteqq \mathbb{R}^{n+1}$ be the standard n-simplex contained in the hyperplane $V^n = \{t = (t_0,\dots,t_n) \mid \sum_i t_i = 0\}$. Consider a C^∞ singular n-simplex $\sigma : \Delta^n \to M$ which extends to an orientation preserving diffeomorphism of a neighbourhood of Δ^n in V^n onto an open set of M. Let $U \subseteqq M$ be the image of $\operatorname{int}\Delta^n$ and let $[\sigma] \in C_n(M)$ denote the n-chain associated to σ. Show that in $C_n(M)$

$$[M] - [\sigma] = \partial c + d$$

for some $c \in C_{n+1}(M)$ and $d \in C_n(M-U)$.

b) For $\omega \in A^n(M)$ recall that the integral $\int_M \omega$ is defined as follows: Choose a finite partition of unity $\{\lambda_\alpha\}$ for M with $\operatorname{supp} \lambda_\alpha \subseteq U_\alpha$ together with orientation preserving diffeomorphisms $\varphi_\alpha : \mathbb{R}^n \to U_\alpha$. Then $\int_M \omega = \sum_\alpha \int_{\mathbb{R}^n} \varphi^*(\lambda_\alpha \omega)$. Show that

$$(7.40) \qquad < I(\omega), [M]> = \int_M \omega.$$

(Hint: First assume ω has support in a set U as considered in a)).

c) Now suppose $n = 2m$ and let $\pi : E \to M$ be a differentiable $SO(2m)$-bundle with connection θ and curvature form Ω. Show using (7.39) that

$$(7.41) \qquad < e(E), [M]> = \int_M \operatorname{Pf}(\Omega^m).$$

In particular for $E = TM$ the tangent bundle of M this proves the Gauss-Bonnet formula

$$(7.42) \qquad \chi(M) = \int_M \operatorname{Pf}(\Omega^m)$$

(in this form due to W. Fenchel and C. B. Allendoerfer - A. Weil).

d) Consider $S^n = \{x \in \mathbb{R}^{n+1} \mid |x| = 1\}$ with the metric induced from $\mathbb{R}^{n+1}$. Observe that $SO(n+1)$ acts on S^n and that if $N = (0,0,\ldots,0,1)$ (the north pole) then the map $\tau : SO(n+1) \to S^n$ given by $g \mapsto gN$ is the principal $SO(n)$-bundle for the tangent bundle of S^n. Consider furthermore the connection in $\tau : SO(n+1) \to S^n$ defined as follows:

For an $(n+1) \times (n+1)$-matric A let $\hat{A}$ denote the $n \times n$ sub-matrix where the last row and column have been cancelled.

Now consider $SO(n+1)$ as a submanifold of $M(n+1,\mathbb{R})$, the set of $(n+1) \times (n+1)$-matrices X. Show that the 1-form

$$\theta = ({}^t X dX)^\wedge \quad ({}^t X = \text{transpose of } X)$$

on $SO(n+1)$ defines a connection.

Now for $n = 2m$ show that

$$\text{(7.43)} \qquad \mathrm{Pf}(\Omega^m) = \frac{(2m)!}{2^{2m}\pi^m m!}\,\nu$$

where ν is the volume form associated with the metric. (Hint: Observe that both sides are invariant under the action of $SO(2m+1)$ so it is enough to evaluate at N. Obs: The volume form has by definition the value $1/(2m)!$ on an orthonormal basis).

Since $\chi(S^{2m}) = 2$ conclude that

$$\text{(7.44)} \qquad vol(S^{2m}) = \frac{2^{2m+1}\pi^m m!}{(2m)!} .$$

8. The Chern-Weil homomorphism for compact groups

In this chapter $H^*(-)$ again means cohomology with real coefficients. The main object is to prove the following

Theorem 8.1. (H. Cartan [8]). Let G be a compact Lie group. Then $w : I^*(G) \to H^*(BG)$ is an isomorphism.

Remark. We shall see below (Proposition 8.3) that for G compact $I^*(G)$ is in principle computable. This also computes $H^*(BG)$ for G any Lie group with a finite number of connected components. In fact in that case G has a maximal compact subgroup K and G/K is diffeomorphic to some Euclidean space (see e.g. G. Hochschild [15, Chapter 15 Theorem 3.1]) so the inclusion $j : K \to G$ induces an isomorphism in homology; hence by Proposition 7.2,

$$Bj^* : H^*(BG) \to H^*(BK)$$

is an isomorphism.

In the following G is a compact Lie group. Let G_0 be the identity component, which is a normal subgroup with G/G_0 the group of components. First let us study $I^*(G)$: In the following we shall identify $I^*(G)$ with the set of invariant polynomial functions, so $P \in I^*(G)$ is now what we denoted by $\tilde{P}$ in Chapter 4 Exercise 1. As mentioned before (cf. Chapter 4 Exercise 4) I^* is a functor so in particular since G acts on G_0 by conjugation we get an induced (right-) action of G on $I^*(G_0)$ by $g \mapsto Ad(g)^*$. By definition G_0 acts trivially, so we have an action of G/G_0 on $I^*(G_0)$ and also by definition

(8.2) $$I^*(G) = \mathrm{Inv}_{G/G_0}(I^*(G_0))$$

the invariant part of $I^*(G_0)$ under the action by G/G_0.

Now suppose G is connected. Then we choose a maximal torus $T \subseteq G$ and consider the Weyl group $W = NT/T$, where NT is the normalizer of T (for the basic properties of maximal tori in compact Lie groups see e.g. J. F. Adams [1, Chapter 4]). Let $i : T \to G$ be the inclusion and let $\mathfrak{g}$ and $\mathfrak{t}$ be the Lie algebras of G and T respectively. Clearly $I^*(T) = S^*(\mathfrak{t}^*)$ and the action of W on $\mathfrak{t}$ induces an action on $I^*(T)$.

Proposition 8.3. Let G be a compact connected Lie group and $i : T \to G$ the inclusion of a maximal torus with Weyl group W. Then i induces an isomorphism

(8.4) $$i^* : I^*(G) \xrightarrow{\cong} \mathrm{Inv}_W(I^*(T)).$$

Proof. If $P \in I^*(G)$ is an invariant polynomial on $\mathfrak{g}$ then clearly the restriction to $\mathfrak{t}$ is invariant under the action by W, so $i^*P \in \mathrm{Inv}_W(I^*(T))$.

i^* injective: Suppose $i^*P = 0$. Every element $v \in \mathfrak{g}$ is contained in a maximal abelian subalgebra and since all such are conjugate (cf. Adams [1, Corollary 4.23]) there is a $g \in G$ such that $\mathrm{Ad}(g)(v) \in \mathfrak{t}$. Hence

$$P(v) = P(\mathrm{Ad}(g)v) = 0,$$

that is, $P = 0$.

i^* surjective: Suppose P is a homogeneous polynomial function of degree k on $\mathfrak{t}$ and suppose P is invariant under W. For $v \in \mathfrak{g}$ choose $g \in G$ such that $\mathrm{Ad}(g)v \in \mathfrak{t}$ and define the function $P' : \mathfrak{g} \to \mathbb{R}$ by

$$P'(v) = P(\mathrm{Ad}(g)v). \tag{8.5}$$

P' is well-defined. In fact suppose

$$t_1 = \mathrm{Ad}(g_1)v, \quad t_2 = \mathrm{Ad}(g_2)v$$

both lie in $\mathfrak{t}$. Then $t_2 = \mathrm{Ad}(g_2 g_1^{-1})t_1$ and then there is an $n \in N(T)$ such that $t_2 = \mathrm{Ad}(n)t_1$ (cf. Adams [1, Lemma 4.33]); hence $P(t_2) = P(t_1)$ since $P \in \mathrm{Inv}_W(I^*(T))$.

We want to show that P' is an invariant polynomial on $\mathfrak{g}$. By definition P' is an <u>invariant</u> function on $\mathfrak{g}$, that is,

$$P'(\mathrm{Ad}(g)v) = P'(v), \quad \forall g \in G, \ v \in \mathfrak{g}, \tag{8.6}$$

and also P' is clearly <u>homogeneous of degree k</u>, that is,

$$P'(\lambda v) = \lambda^k P'(v), \quad \forall v \in \mathfrak{g}, \ \lambda \in \mathbb{R}. \tag{8.7}$$

In an appendix to this chapter we shall show that P' is a C^∞ function on $\mathfrak{g}$ (a surprisingly non-trivial fact). Then P' is actually a homogeneous polynomial of degree k due to the following lemma:

<u>Lemma 8.8</u>. Suppose $f : \mathbb{R}^n \to \mathbb{R}$ is a C^∞ function which is homogeneous of degree k, that is satisfies

$$f(\lambda x) = \lambda^k f(x), \quad \forall x \in \mathbb{R}^n, \ \lambda \in \mathbb{R}. \tag{8.9}$$

Then f is a homogeneous polynomial of degree k.

<u>Proof</u>. Let $x = (x_1,\dots,x_n)$ be the coordinates in $\mathbb{R}^n$. Differentiating (8.9) k times with respect to λ using the chain rule and putting $\lambda = 0$ yields

$$(8.10) \qquad \sum_{i_1+\dots+i_n=k} a_{i_1\dots i_n} x_1^{i_1}\dots x_n^{i_n} = k!f(x), \quad x \in \mathbb{R}^n,$$

where

$$a_{i_1\dots i_n} = \frac{\partial^k f}{\partial x_1^{i_1}\dots\partial x_n^{i_n}}(0).$$

This proves the lemma and ends the proof of the proposition.

In view of Proposition 8.3 we shall first prove Theorem 8.1 for $G = T^n$ the <u>n-dimensional</u> torus, i.e.

$$T^n = T^1 \times\dots\times T^1 \quad (n \text{ times})$$

where $T^1 = U(1)$ is the unit circle group in $\mathbb{C}$. We shall identify $T^n = \mathbb{R}^n/\mathbb{Z}^n$ via the map

$$\exp(x_1,\dots,x_n) = (e^{2\pi i x_1},\dots,e^{2\pi i x_n}), \quad (x_1,\dots,x_n) \in \mathbb{R}^n.$$

Then the Lie algebra of T^n is $\mathfrak{t}^n = \mathbb{R}^n$ with zero Lie bracket, so $I^*(T^n) = S^*((\mathbb{R}^n)^*)$ is actually identified with the polynomial ring in the variables $x_1,\dots,x_n$. For $n = 1$ $\mathfrak{t}^1$ is identified with $\mathfrak{u}(1) = i\mathbb{R} \subseteq \mathbb{C}$ under the map $x \mapsto 2\pi i x$ and it follows that $I^*(T^1)$ is the polynomial ring in one variable x with Chern-Weil image $w(x) = -c_1 \in H^2(BT^1)$.

<u>Proposition 8.11</u>. $H^*(BT^1)$ is a polynomial ring in the variable $w(x) \in H^2(BT^1)$ where x is the identity polynomial on $\mathfrak{t}^1 = \mathbb{R}$.

<u>Proof</u>. By Proposition 6.1, $H^*(BT^1)$ can be calculated as the homology of the total complex of the double complex $A^{p,q}(NT^1)$ with

$$A^{p,q}(NT^1) = A^q(NT^1(p)) = A^q(T^p)$$

As above identify $T^p = \mathbb{R}^p/\mathbb{Z}^p$ with coordinates $(x_1,\dots x_p)$.

Now consider the double complex

$$A_0^{p,q} \subseteq A^{p,q}(NT^1) \tag{8.12}$$

where $A_0^{p,q}$ is the vectorspace spanned by all $dx_{j_1} \wedge \ldots \wedge dx_{j_q}$, $1 \leqq j_1 < \ldots < j_q \leqq p$ (so $A_0^{p,q} \cong \Lambda^q(\mathbb{R}^p)$). Notice that $A_0^{p,q} = 0$ for $p < q$ and that the vertical differential d'' vanishes on $A_0^{p,q}$. It is easy to see that the inclusion (8.12) induces an isomorphism

$$A_0^{p,q} \xrightarrow{\cong} H^q(A^{p,*}(NT^1)).$$

Hence by Lemma 1.19 the inclusion (8.12) induces an isomorphism on homology of the total complexes. It follows that

$$H^n(BT^1) \cong \coprod_{p+q=n} H^p(A_0^{*,q}) \tag{8.13}$$

so we shall calculate $H^p(A_0^{*,q})$ for each q. Here $\varepsilon_i : T^{p+1} \to T^p$, $i = 0,\ldots,p$, is given by

$$\varepsilon_i(x_1,\ldots,x_{p+1}) = \begin{cases} (x_2,\ldots,x_{p+1}), & i = 0, \\ (x_1,\ldots,x_i+x_{i+1},\ldots,x_{p+1}), & i = 1,\ldots,p, \\ (x_1,\ldots,x_p), & i = p+1, \end{cases}$$

$$(x_1,\ldots,x_{p+1}) \in \mathbb{R}^{p+1}/\mathbb{Z}^{p+1}.$$

By a straightforward calculation it is seen that

$$\delta' = \sum_{i=0}^{p+1} (-1)^i \varepsilon_i^* : A_0^{p,q} \to A_0^{p+1,q}$$

is given by

$$\delta'(dx_{j_1} \wedge \ldots \wedge dx_{j_q}) = \Big(\sum_{i=0}^{j_1} (-1)^i\Big) dx_{j_1+1} \wedge \ldots \wedge dx_{j_q+1} + \tag{8.14}$$

$$+ \Big(\sum_{i=j_1}^{j_2} (-1)^i\Big) dx_{j_1} \wedge dx_{j_2+1} \wedge \ldots \wedge dx_{j_q+1} + \ldots +$$

$$+ \Big(\sum_{i=j_q}^{p+1} (-1)^i\Big) dx_{j_1} \wedge \ldots \wedge dx_{j_q}.$$

Now define maps

$$R : A_0^{p-1,q-1} \overset{\leftarrow}{\rightarrow} A_0^{p,q} : T$$

for $q = 1,2,\ldots,$ $p = q,q+1,\ldots,$ by

(8.15) $$R(\omega) = \omega \wedge dx_p, \quad \omega \in A_0^{p-1,q-1}$$

(8.16) $$T(\alpha \wedge dx_p + \beta) = \alpha, \quad \alpha \wedge dx_p + \beta \in A^{p,q}$$

α, β do <u>not</u> contain dx_p.

Then it is easily checked that R and T give chain maps between the complexes $A_0^{*-1,q-1}$ and $A_0^{*,q}$, and clearly

(8.17) $$T \circ R = \text{id}.$$

On the other hand if we let $s_p : T^{p-1} \to T^p$, $p = 1,2,\ldots,$ be induced by

$$s_p(x_1,\ldots,x_{p-1}) = (x_1,\ldots,x_{p-1},0), \quad (x_1,\ldots,x_{p-1}) \in \mathbb{R}^{p-1},$$

then it is easy to check that

(8.18) $$(-1)^p \delta \circ s_p^* + (-1)^{p+1} s_{p+1} \circ \delta = \text{id} - R \circ T$$

on $A_0^{p,q}$. The details are left as an exercise. It follows that

(8.19) $$H^p(A_0^{0,q}) \cong H^{p-q}(A_0^{*,0}) = \begin{cases} 0 & p \neq q, \\ \mathbb{R} & p = q. \end{cases}$$

Also the generator is represented by 1 for $p = q = 0$ and $dx_1 \wedge \ldots \wedge dx_p$ for $p = q > 0$. By (8.13) we now have

$$H^n(BT^1) = \begin{cases} 0 & n \text{ odd} \\ \mathbb{R} & n \text{ even} \end{cases}$$

and we want to show that $\omega(x)^p \neq 0$. For this notice that $\omega(x)$ is represented in $A^2(NT^1)$ by the curvature form Ω of the connection θ given in $N\bar{T}^1$ by

$$\theta = \sum_{i=0}^{p} t_i dy_i \quad \text{on} \quad \Delta^p \times N\bar{T}^1(p)$$

where $N\bar{T}^1(p) = T^{p+1} = \mathbb{R}^{p+1}/\mathbb{Z}^{p+1}$ has the coordinates $(y_0,\ldots,y_p)$. Now

$$\Omega = d\theta = \sum_{i=0}^{p} dt_i \wedge dy_i = \sum_{i=1}^{p} dt_i \wedge (dy_i - dy_0).$$

It follows that on $\Delta^p \times N\bar{T}^1(p)$

$$\Omega^p = \pm p!dt_1 \wedge \ldots \wedge dt_p \wedge (dy_1 - dy_0) \wedge \ldots \wedge (dy_p - dy_0),$$

which is the lift of

$$\Omega^p = \pm p!dt_1 \wedge \ldots \wedge dt_p \wedge dx_1 \wedge \ldots \wedge dx_p$$

on $\Delta^p \times NT^1(p)$ since $dx_i = dy_{i-1} - dy_i$, $i = 1,\ldots,p$, by (5.8). Therefore $\omega(x)^p$ is represented in $A^{p,p}(NT^1)$ by $I_\Delta(\Omega^p) = \pm dx_1 \wedge \ldots \wedge dx_p$ which represents $\pm$ the generator in cohomology. This ends the proof of the proposition.

Remark. Notice that $SO(2) = T^1$ and that the classes defined in Chapter 7 e and $\mathcal{e}$ are both identified in $H^2(BSO(2),\mathbb{R})$ with $-\omega(x) \in H^2(BT^1)$ by (7.25) and (7.38). It now follows from (7.24) and (7.36) that when a $SO(2m)$-bundle $\pi : E \to X$ is the Whitney sum of m $SO(2)$-bundles then $e(E) = \mathcal{e}(E)$ in $H^{2m}(X,\mathbb{R})$.

To prove Theorem 8.1 for $G = T^n$ we now need the following proposition:

Proposition 8.20. a) For any Lie group G, the space EG is contractible. In particular there is a natural isomorphism of homotopy groups

(8.21) $$\pi_i(BG) \cong \pi_{i-1}(G), \quad i = 1,2,\ldots$$

b) For G and H any Lie groups the natural map

$$Bp_1 \times Bp_2 : B(G \times H) \to BG \times BH$$

induced by the projections $p_1 : G \times H \to G$, $p_2 : G \times H \to H$, is a weak homotopy equivalence, in particular it induces an isomorphism in cohomology with any coefficients.

Proof. a) By definition EG is a quotient space of $\coprod_{p=0}^{\infty} \Delta^p \times N\bar{G}(p)$. Define the homotopy $h_s : EG \to EG$, $s \in [0,1]$, by

$$h_s((t_0,t_1,\ldots,t_p),(g_0,\ldots,g_p)) =$$

$$= ((1-s,st_0,\ldots,st_p),(1,g_0,\ldots,g_p)).$$

This is easily seen to be well-defined and a contraction of EG to the point $((1),(1)) \in \Delta^0 \times G$. (8.21) now clearly follows from the homotopy sequence for the fibration $EG \to BG$ with fibre G.

b) p_1 and p_2 clearly induce a map of principal $G \times H$-bundles

$$\begin{array}{ccc} E(G \times H) & \to & EG \times EH \\ \downarrow & & \downarrow \\ B(G \times H) & \to & BG \times BH. \end{array}$$

Since both total spaces are contractible the map $B(G \times H) \to BG \times BH$ induces an isomorphism on homotopy groups by (8.21), and the second statement follows from Whitehead's theorem (see e.g. E. Spanier [28, Chapter 7 § 5, Theorem 9]).

Corollary 8.22. $w : I^*(T^n) \to H^*(BT^n)$ is an isomorphism. That is, $H^*(BT^n)$ is a polynomial ring in the variables

$w(x_i) \in H^2(BT^n)$, $i = 1,\ldots,n$, where x_i are the canonical generators of $I^*(T^n) = S^*((\mathbb{R}^n)^*) = \mathbb{R}[x_1,\ldots,x_n]$.

Proof. Obvious from the Propositions 8.11 and 8.20 together with the Künneth theorem.

Before we proceed with the proof of Theorem 8.1 we need a few preparations of a technical nature: Let M be an n-dimensional manifold and let N be a compact oriented k-dimensional manifold. Consider the projection $p : M \times N \to M$. We shall use a homomorphism

$$p_* : A^*(M \times N) \to A^{*-k}(M)$$

called integration along N. The reader will recognize the technique from the Poincaré lemma (Lemma 1.2). First suppose that $\omega \in A^l(M \times N)$, $l \geq k$, has support inside $M \times U$, where U is a coordinate neighbourhood in N with local coordinates $(x_1,\ldots,x_k)$. Suppose furthermore that the coordinates are chosen such that $dx_1 \wedge \ldots \wedge dx_k$ is a positive k-form with respect to the orientation of N. Then we can write

(8.23) $\omega = dx_1 \wedge \ldots \wedge dx_k \wedge \alpha + \beta$ on $M \times U$,

where α does not contain $dx_1,\ldots,dx_k$ and β only involves terms containing $dx_1,\ldots,dx_k$ to a degree less than k. Then we define $p_*\omega \in A^{l-k}(M)$ to be

(8.24) $p_*\omega = \displaystyle\int_U dx_1 \wedge \ldots \wedge dx_k \wedge \alpha,$

which means that we integrate the coefficients of α as functions of $x_1,\ldots,x_k$. We leave it as an exercise to verify that $p_*\omega$ is well-defined and that the definition of $p_*\omega$ extends to all forms on $M \times N$ using a partition of unity.

Also the following lemma is left as an exercise:

Lemma 8.25. a) $p_*(d\omega) = (-1)^k dp_*\omega$, $\omega \in A^*(M \times N)$.

b) For $\omega \in A^*(M \times N)$ and $\mu \in A^*(M)$

$$p_*(\omega \wedge p^*\mu) = (p_*\omega) \wedge \mu.$$

Proof of Theorem 8.1. First assume G connected and choose a maximal compact subgroup T. Let $i : T \to G$ be the inclusion and consider the commutative diagram

(8.26)
$$
\begin{array}{ccc}
I^*(G) & \xrightarrow{\;i^*\;} & I^*(T) \\
{\scriptstyle w}\Big\downarrow & & \Big\downarrow{\scriptstyle w} \\
H^*(BG) & \xrightarrow{\;Bi^*\;} & H^*(BT).
\end{array}
$$

First notice that by functoriality $W = NT/T$ acts on $H^*(BT)$ and the image of Bi^* is contained in the invariant part. In fact for $g \in G$ let $\iota_g : G \to G$ be the inner conjugation $x \mapsto g^{-1}xg$. Then by Lemma 5.20 $N\iota_g : NG \to NG$ is simplicially homotopic to the identity so by Corollary 5.17, $B\iota_g : BG \to BG$ induces the identity on cohomology. Also by functoriality $w : I^*(T) \to H^*(BT)$ is a map of W-modules, hence (8.26) yields the commutative diagram

(8.27)
$$
\begin{array}{ccc}
I^*(G) & \xrightarrow[\;\cong\;]{\;i^*\;} & \mathrm{Inv}_W(I^*(T)) \\
{\scriptstyle w}\Big\downarrow & & \cong\Big\downarrow{\scriptstyle w} \\
H^*(BG) & \xrightarrow{\;Bi^*\;} & \mathrm{Inv}_W(H^*(BT))
\end{array}
$$

where the upper horizontal map and right vertical map are isomorphisms by Proposition 8.3 and Corollary 8.22, respectively. Therefore it is enough to show

(8.28) $\qquad\qquad Bi^* : H^*(BG) \to H^*(BT)$ is injective.

To prove this we consider the commutative diagram

$$\begin{array}{ccc} ET/T & \longrightarrow & EG/T \\ \cong\downarrow & & \downarrow \\ BT & \xrightarrow{Bi} & BG \end{array}$$

and observe that the upper horizontal map is a weak homotopy equivalence by Proposition 8.20 a). Hence (8.28) is equivalent to show that the map $EG/T \to BG$ induces an injective map in cohomology. This map is the realization of the map of simplicial manifolds $i' : N\bar{G}/T \to NG$ induced by the map $\gamma : N\bar{G} \to NG$ given by (5.8). Here $N\bar{G}/T$ is the simplicial manifold with

$$(N\bar{G}/T)(p) = N\bar{G}(p)/T,$$

where T acts by the diagonal action on the right of $N\bar{G}(p) = G \times \ldots \times G$ (p+1 times). This, however, can be identified with the simplicial manifold $N(G;G/T)$ where

$$N(G;G/T)(p) = NG(p) \times G/T$$

and $\varepsilon_i : N(G;G/T)(p) \to (NG/T)(p-1)$, $i = 0,\ldots,p$, is given by

$$\varepsilon_i(g_1,\ldots,g_p,gT) = \begin{cases} (g_2,\ldots,g_p,gT), & i = 0, \\ (g_1,\ldots,g_ig_{i+1},\ldots,g_p,gT), & i = 1,\ldots,p-1 \\ (g_1,\ldots,g_{p-1},g_pgT), & i = p. \end{cases}$$

In fact the identification

$$(8.29) \qquad N(G;G/T) \approx N\bar{G}/T$$

is given by the map

$$(g_1,\ldots,g_p,g\cdot T) \to (g_1\cdots g_p\cdot g,\ldots,g_p\cdot g,g)T.$$

Under this identification the map $i' : N\bar{G}/T \to NG$ corresponds to the map $\tilde{i} : N(G;G/T) \to NG$ given by the projection on the first factor in $NG(p) \times G/T$. (8.28) is therefore equivalent to

(8.28)' $\|\tilde{i}\|^* : H^*(\|NG\|) \to H^*(\|N(G;G/T)\|)$ is injective,

which is proved as follows:

We shall see below that G/T is an orientable manifold of even dimension, say $2m$, and (8.24) therefore produces a map

$$\tilde{i}_* : A^*(\Delta^p \times NG(p) \times G/T) \to A^{*-2m}(\Delta^p \times NG(p))$$

for each p. It is easy to see that these maps preserve the requirements in Definition 6.2 so we get a map

$$\tilde{i}_* : A^*(N(G;G/T)) \to A^{*-2m}(NG).$$

Now suppose we have proved the following

<u>Lemma 8.30</u>. There is a 2m-form $\Psi \in A^{2m}(N(G;G/T))$ such that

(i) $d\Psi = 0$

(ii) The restriction Ψ^0 to $G/T = \Delta^0 \times N(G;G/T))(0)$ satisfies $\int_{G/T} \Psi^0 \neq 0$.

We can then define a map

$$\tau : A^*(N(G;G/T)) \to A^*(NG)$$

by

$$\tau(\varphi) = \tilde{i}_*(\Psi \wedge \varphi), \quad \varphi \in A^*(N(G;G/T)).$$

By Lemma 8.25 a) and Lemma 8.30 (i), τ is a chain map, hence induces a map in cohomology

$$\tau_* : H^*(\| N(G; G/T) \|) \to H^*(\| NG \|).$$

By Lemma 8.25 b),

$$\tau \circ \tilde{i}^*(\varphi) = \tilde{i}_*(\Psi) \cdot \varphi, \quad \varphi \in A^*(NG).$$

Here $\tilde{i}_*(\Psi) \in A^0(NG)$ is closed, hence a constant, that is,

$$\tilde{i}^*(\Psi) = \int_{G/T} \Psi^0 \neq 0.$$

It follows that $\tau \circ \tilde{i}^*$ and so also $\tau_* \circ \|\tilde{i}\|^*$: $H^*(\| NG \|) \to H^*(\| NG \|)$ is multiplication by a non-zero constant. This shows the injectivity of $\|\tilde{i}\|^*$ and hence proves (8.28). It remains to prove the existence of Ψ:

Proof of Lemma 8.30. Choose an inner product on $\mathfrak{g}$ which is invariant under the adjoint action of G. (This is possible since G is compact). Now make a root space decomposition of $\mathfrak{g}$, that is, split $\mathfrak{g}$ into an orthogonal direct sum

$$\mathfrak{g} = \mathfrak{t} \oplus \mathfrak{m}$$

and find an orthonormal basis $\{e_1, \ldots, e_{2m}\}$ for $\mathfrak{m}$ such that $\mathrm{Ad}(\exp(t))$, $t \in \mathfrak{t}$ acts on $\mathfrak{m}$ by the matrix

(8.31)
$$\mathrm{Ad}(\exp(t)) = \begin{pmatrix} \cos 2\pi\alpha_1(t) & -\sin 2\pi\alpha_1(t) & & & \\ \sin 2\pi\alpha_1(t) & \cos 2\pi\alpha_1(t) & & 0 & \\ & & \ddots & & \\ & 0 & & \cos 2\pi\alpha_m(t) & -\sin 2\pi\alpha_m(t) \\ & & & \sin 2\pi\alpha_m(t) & \cos 2\pi\alpha_m(t) \end{pmatrix}$$

where $\alpha_i : \mathfrak{t} \to \mathbb{R}$, $i = 1, \ldots, m$, are linear forms on $\mathfrak{t}$ (for details see e.g. Adams [1, Chapter 4]). Notice that the tangent bundle of G/T can be identified with the 2m-dimensional vector bundle

$$\xi : G \times_T \mathfrak{m} \to G/T$$

which is clearly an oriented bundle with the orientation given by the basis $\{e_1,\ldots,e_{2m}\}$.

Now let $\kappa : \mathfrak{g} \to \mathfrak{t}$ be the orthogonal projection and let θ be the canonical connection in $N\bar{G}$ given by (6.12). Then clearly

$$\theta_T = \kappa \circ \theta$$

defines a connection in the principal T-bundle $N\bar{G} \to N\bar{G}/T$ and let Ω_T be the curvature form. Also consider $P \in I^m(T)$ given by the polynomial function

$$P(v,\ldots,v) = (-1)^m \prod_{i=1}^{m} \alpha_i(v), \quad v \in \mathfrak{t} .$$

Then by Chern-Weil theory the 2m-form $P(\Omega_T^m)$ is a closed form on $N\bar{G}/T$ and we let Ψ be the corresponding form on $N(G;G/T)$ under the identification (8.29), so clearly $d\Psi = 0$ is satisfied. It remains to prove (ii). Now $\Psi^0 = P(\Omega_T^m) \in A^{2m}(G/T)$ is just the Chern-Weil image of P in the principal T-bundle $G \to G/T$ with connection θ_T given by

$$(\theta_T)_g = \kappa \circ (L_{g^{-1}})_*, \quad g \in G,$$

and $\Omega_T = d\theta_T$. Unfortunately it is not so easy to calculate $\int_{G/T} P(\Omega_T^m)$ directly. However, as noticed above the extension of the bundle $G \to G/T$ to the group $SO(2m)$ via the adjoint representation on $\mathfrak{m}$ is just the tangent bundle of G/T and it is easy to see that $P(\Omega_T^m)$ is exactly the Pfaffian form. On the other hand it follows from (8.31) that the bundle is a Whitney sum of $SO(2)$-bundles. Therefore, as remarked after

Proposition 8.11 (cf. Exercise 2 of Chapter 7),

$$(8.32) \qquad \int_{G/T} P(\Omega_T^m) = \langle e(T(G/T)),[G/T]\rangle.$$

Now the right hand side of (8.32) we can compute by the formula (7.33) for a vector field of the following form: Choose a regular element $v_0 \in \mathfrak{t}$ (i.e. $\alpha_i(v_0) \neq 0$ for every root α_i, $i = 1,\dots,m$) and consider the section s of the vector bundle $\xi : G \times_T \mathfrak{m} \to G/T$ given by

$$s(gT) = (g,(\mathrm{id}-\kappa) \circ \mathrm{Ad}(g^{-1})v_0), \quad g \in G$$

where again $\kappa : \mathfrak{g} \to \mathfrak{t}$ is the orthogonal projection. Since v_0 is regular $s(gT) = 0$ iff $g \in NT$ so s vanishes at the finite set of points $W = NT/T \subseteq G/T$. Now we claim that the local index of s at $gT \in W$ is $+1$. For this we recall the well-known fact that the exponential map $\exp : \mathfrak{m} \to G \to G/T$ maps a neighbourhood of $0 \in \mathfrak{m}$ diffeomorphic onto a neighbourhood of $\{T\}$ in G/T, so we get a local trivialization near gT by

$$(g \exp x,v) \mapsto v, \quad x \in \mathfrak{m} \text{ near zero, } v \in \mathfrak{m}.$$

It is therefore enough to see that the map $\tilde{s} : \mathfrak{m} \to \mathfrak{m}$ given by

$$\tilde{s}(x) = (\mathrm{id}-\kappa)(\mathrm{Ad}(\exp(-x))v_g), \quad v_g = \mathrm{Ad}(g^{-1})v, \; x \in \mathfrak{m},$$

is an orientation preserving diffeomorphism near 0. The differential $\tilde{s}_*$ at 0 is given by $\tilde{s}_*(x) = -[x,v_g] = \mathrm{ad}(v_g)(x)$, $x \in \mathfrak{m}$. Differentiating (8.31) and taking the determinant now gives

$$\det(\mathrm{ad}\, v_g) = (2\pi)^m \prod_{i=1}^m \alpha_i(v_g)^2 > 0$$

so the local index of s at gT is +1. It follows that

$$\int_{G/T} \Psi^0 = \int_{G/T} P(\Omega_T^m) = |W| > 0$$

which proves Lemma 8.30 and finishes the proof of Theorem 8.1 for G connected.

For G a general compact group we get a diagram similar to (8.27):

$$\begin{array}{ccc} I^*(G) & \xrightarrow{\cong} & \mathrm{Inv}_{G/G_0}(I^*(G_0)) \\ \downarrow & & \downarrow \cong \\ H^*(BG) & \longrightarrow & \mathrm{Inv}_{G/G_0}(H^*(BG_0)) \end{array} \tag{8.33}$$

where again the upper horizontal map is the isomorphism (8.2) and the right vertical map is an isomorphism since G_0 is connected. Again it suffices to show that if $i : G_0 \to G$ is the inclusion then

(8.34) $\qquad Bi^*: H^*(BG) \to H^*(BG_0)$ is injective.

As before, this is equivalent to showing that $\|\tilde{i}\|^* : H^*(\|NG\|) \to H^*(\|N(G;G/G_0)\|)$ is injective, where $\tilde{i} : N(G;G/G_0) \to NG$ is defined as follows:

$$N(G;G/G_0)(p) = NG(p) \times G/G_0$$

and $\tilde{i} : N(G;G/G_0)(p) \to NG(p)$ is given by the projection on the first factor. This time

$$\tau : \quad A^*N(G;(G/G_0))) \to A^*(NG)$$

is simply given by

$$\tau(\varphi) = \sum_{g\in G/G_0} s_g^*\varphi, \qquad \varphi \in A^*(N(G;G/G_0))$$

where $s_g : \Delta^p \times NG(p) \to \Delta^p \times N(G;G/G_0)(p)$ is given by putting

$gG_0 \in G/G_0$ on the last coordinate (notice that s_g is not a simplicial map but still τ is well-defined). Again it is easily checked that τ is a chain map and that $\tau \circ \tilde{I}^*$ is multiplication by $|G/G_0|$. Hence also $\tau_* \circ \|\tilde{I}\|^*$ is multiplication by $|G/G_0|$ where

$$\tau_* : H^*(\| N(G;G/G_0)\|) \to H^*(\| NG \|)$$

is the map induced by τ. This shows (8.34) and ends the proof of Theorem 8.1.

Corollary 8.35. (A. Borel [3]). Let G be a compact connected Lie group, and let $i : T \to G$ be the inclusion of a maximal torus. Then $Bi : BT \to BG$ induces an isomorphism

$$H^*(BG) \xrightarrow{\cong} \mathrm{Inv}_W H^*(BT).$$

Proof. Obvious from the diagram (8.27).

Corollary 8.36. (i) The Chern classes of $Gl(n,\mathbb{C})$-bundles are uniquely determined by the properties a) - d) of Theorem 7.3.

(ii) Furthermore

$$H^*(BGl(n,\mathbb{C})) \cong H^*(BU(n)) \cong \mathbb{R}[c_1,\dots,c_n]$$

is a polynomial ring with the Chern classes $c_1,\dots,c_n$ as generators.

Proof. As noticed in Chapter 7 it is enough to consider $U(n)$-bundles. Now let $i : T^n \to U(n)$ be the natural inclusion

$$i(\lambda_1,\dots,\lambda_n) = \begin{pmatrix} \lambda_1 & & & 0 \\ & \lambda_2 & & \\ & & \ddots & \\ 0 & & & \lambda_n \end{pmatrix} \quad \lambda_1,\dots,\lambda_n \in U(1)$$

and let $q_j : T^n \to U(1)$ be the projection onto the j-th factor, $j = 1,\dots,n$. It is well-known that T^n is a maximal torus so by Corollary 8.35

$$Bi^* : H^*(BU(n)) \to H^*(BT^n)$$

is injective. That is, the Chern classes are determined by the values on $U(n)$-bundles which are Whitney sums of $U(1)$-bundles. Hence by (7.5) they are determined by c_1 on $U(1)$-bundles. This, however, is determined by (7.6) as remarked immediately after Proposition 8.11. This proves (i).

(ii) By Corollary 8.22 $H^*(BT^n) = \mathbb{R}[y_1,\dots,y_n]$ where $y_j = (Bq_j)^*c_1 \in H^2(BT^n)$, $j = 1,\dots,n$ and $c_1 \in H^2(BU(1))$ is the first Chern class. Now W is the symmetric group acting on T^n by permuting the factors, i.e. W acts on $H^*(BT^n)$ by permuting $y_1,\dots,y_n$. Hence $\mathrm{Inv}_W(BT^n)$ is a polynomial ring with generators the <u>elementary symmetric polynomials</u> $\sigma_k(y_1,\dots,y_n)$, $k = 1,\dots,n$, in $y_1,\dots,y_n$ (see e.g. B.L. van der Waerden [32, § 29]). However, by (7.5)

$$\sigma_k(y_1,\dots,y_n) = i^*c_k, \quad k = 1,\dots,n,$$

which proves the corollary.

<u>Corollary 8.37</u>. (i) The Euler class with real coefficients for $SO(2m)$-bundles is uniquely determined by the properties i), ii), and v) of Exercise 1 e) of Chapter 7. In particular formula (7.39) holds.

(ii) Furthermore

$$H^*(BGl(2m,\mathbb{R})^+) \cong H^*(BSO(2m))$$
$$\cong \mathbb{R}[p_1,\dots,p_{m-1},e]$$

is a polynomial ring with generators the first $m-1$ Pontrjagin classes $p_1,\dots,p_{m-1}$ and the Euler class e.

(iii) Finally

$$H^*(BGl(2m,\mathbb{R})) \cong H^*(BO(2m)) \cong \mathbb{R}[p_1,\dots,p_m]$$

is a polynomial ring in the Pontrjagin classes $p_1,\dots,p_m$.

Proof. The maximal torus in $SO(2m)$ is well-known to be the set T^m of matrices of the form

$$\begin{pmatrix} \cos 2\pi x_1 & -\sin 2\pi x_1 & & & \\ \sin 2\pi x_1 & \cos 2\pi x_1 & & & 0 \\ & & \ddots & & \\ & 0 & & \cos 2\pi x_m & -\sin 2\pi x_m \\ & & & \sin 2\pi x_m & \cos 2\pi x_m \end{pmatrix}$$

$(x_1,\dots,x_m) \in \mathbb{R}^m/\mathbb{Z}^m$. Again let $i : T^m \to SO(2m)$ be the inclusion and let $q_j : T^m \to SO(2)$, $j = 1,\dots,m$, be the projection on the j-th factor. As before (i) follows from the injectivity of $i^* : H^*(BSO(m)) \to H^*(BT^m)$ together with the remark following Proposition 8.11.

(ii) Again $H^*(BT^m) \cong \mathbb{R}[y_1,\dots,y_m]$ where $y_j = (Bq_j)^*e \in H^2(BT^m)$. It is easily seen (cf. Adams Example 5.17) that the Weyl group W acts on $H^*(BT^m)$ by permuting the y_j's and changing the sign on an even number of the y_j's. We want to determine the subring

$$A = \mathrm{Inv}_W(\mathbb{R}[y_1,\dots,y_m]) \subseteq \mathbb{R}[y_1,\dots,y_m].$$

First notice that A has an involution $\tau : A \to A$ given by changing the sign of y_1, say. Then clearly $A = A_+ \oplus A_-$, where A_+ and A_- are the ± 1 eigen spaces for τ. Notice that

$$A_+ = \mathrm{Inv}_{W'}(\mathbb{R}[y_1,\ldots,y_m])$$

where W' is the group generated by the permutations of the y_j's together with the transformations which changes the sign of any number of the y_j's. It is now easily seen that

$$A_+ = \mathbb{R}[\sigma_1,\ldots,\sigma_m]$$

where $\sigma_j = \sigma_j(y_1^2,\ldots,y_m^2)$ is the j-th elementary symmetric polynomial in $y_1^2,\ldots,y_m^2$. Now every element of A_- is easily seen to be divisible by the polynomial

$$\varepsilon = y_1\ldots y_m.$$

Hence

$$A = A_+ \oplus A_+\varepsilon.$$

Now $\varepsilon^2 = \sigma_m(y_1^2,\ldots,y_m^2) \in A_+$; hence

$$A = \mathbb{R}[\sigma_1,\ldots,\sigma_{m-1},\varepsilon].$$

Here $\sigma_j = (Bi)^*p_j$, $j = 1,\ldots,m$, by (7.20) and (7.26), and $\varepsilon = (Bi)^*e$ by (7.24). This proves (ii).

(iii) By Theorem 8.1 and (8.2)

$$\begin{aligned} H^*(BO(2m)) &\cong I^*(O(2m)) \\ &\cong \mathrm{Inv}_{O(2m)/SO(2m)}(I^*(SO(2m))). \end{aligned}$$

Here $O(2m)/SO(2m) \cong \mathbb{Z}/2$ acts on $I^*(SO(2m))$ using the adjoint action of an orientation reversing orthogonal matrix. This clearly fixes the Pontrjagin polynomials and changes the sign of the Pfaffian polynomial (see Chapter 4, Example 1 and 3). Hence the invariant part of $I^*(SO(2m))$ is the polynomial ring in the variables $P_1,\dots,P_{m-1}$ and $Pf^2 = P_m$. This proves the corollary.

In a similar way one proves

Corollary 8.38.

(i) $H^*(BGl(2m+1)^+) \cong H^*(BSO(2m+1)) \cong \mathbb{R}[p_1,\dots,p_m]$

is a polynomial ring in the Pontrjagin classes.

(ii) $H^*(BGl(2m+1)) \cong H^*(BO(2m+1)) \cong H^*(BSO(2m+1))$

$$\cong \mathbb{R}[p_1,\dots,p_m].$$

Remark. In all the cases considered above $H^*(BG) = \mathrm{Inv}_W(S^*(\mathfrak{t}^*))$ is a polynomial ring. This is no coincidence. In fact if V is any real vectorspace of dimension l and W is a finite group generated by reflections in hyperplanes of V, then $\mathrm{Inv}_W(S^*(V^*))$ is a polynomial ring in l generators (cf. N. Bourbaki [6, Chapitre V, § 5, théorème 3]).

APPENDIX

We will in this appendix give a proof of the differentiability of the function $P' : \mathfrak{g} \to \mathbb{R}$ defined in the proof of Proposition 8.3 by the formula (8.5). First we recall some rather standard facts from the theory of Lie groups.

In the following suppose G is a compact connected semi-simple Lie group without center. Let $\mathfrak{g}$ be the Lie algebra and $\mathfrak{g}_{\mathbb{C}} = \mathfrak{g} \otimes_{\mathbb{R}} \mathbb{C}$ the complexification of $\mathfrak{g}$. Then there is a complex analytic Lie group $G_{\mathbb{C}}$ (the complexification of G)

and an injection $j : G \to G_{\mathbb{C}}$ such that $\mathfrak{g}_{\mathbb{C}}$ is the Lie algebra of $G_{\mathbb{C}}$ and $j_* : \mathfrak{g} \to \mathfrak{g}_{\mathbb{C}}$ is the natural inclusion $\mathfrak{g} \to \mathfrak{g} \oplus i\mathfrak{g} = \mathfrak{g}_{\mathbb{C}}$. To see this notice that since G is without center $Ad : G \to Gl(\mathfrak{g})$ is injective and the image is the connected subgroup $Int(\mathfrak{g}) \subseteq Gl(\mathfrak{g})$ with Lie algebra $ad(\mathfrak{g}) \subseteq End(\mathfrak{g})$ defined by

$$ad(\mathfrak{g}) = \{ad(v) \mid v \in \mathfrak{g}\ ,\ ad(v)(x) = [v,x],\ x \in \mathfrak{g}\}.$$

We can then take $G_{\mathbb{C}} = Int(\mathfrak{g}_{\mathbb{C}}) \subseteq Gl(\mathfrak{g}_{\mathbb{C}})$ the complex analytic group with complex Lie algebra $ad(\mathfrak{g}_{\mathbb{C}}) \subseteq End(\mathfrak{g}_{\mathbb{C}})$. Here again $ad : \mathfrak{g}_{\mathbb{C}} \to ad(\mathfrak{g}_{\mathbb{C}})$ is an isomorphism and $j : G \to G_{\mathbb{C}}$ is given by the composite

$$G \xrightarrow{Ad} Int(\mathfrak{g}) \longrightarrow Int(\mathfrak{g}_{\mathbb{C}}).$$

In the following we shall identify G with the image in $G_{\mathbb{C}}$.

We also need the <u>Jordan-decomposition</u> of elements of $\mathfrak{g}_{\mathbb{C}}$:

For a complex vector space V a linear map $A \in End(V)$ has a unique <u>Jordan-decomposition</u>

$$A = S + N, \quad SN = NS$$

with S <u>semi-simple</u> (i.e. V has a basis of eigenvectors for S) and N <u>nilpotent</u> (i.e. $N^k = 0$ for some $k \geq 0$). In particular for $v \in \mathfrak{g}_{\mathbb{C}}$ we have a Jordan-decomposition of $ad(v) \in End(\mathfrak{g}_{\mathbb{C}})$ and we have

<u>Lemma 8.A.1</u>. For $v \in \mathfrak{g}_{\mathbb{C}}$ there is a unique Jordan-decomposition $v = s + n$ such that $ad\,v$ is semi-simple, $ad\,n$ is nilpotent and $[s,n] = 0$

<u>Proof</u>. We must show that the <u>semi-simple</u> part of $ad\,v$ (and hence also the nilpotent part) lies again in

$\mathrm{ad}(\mathfrak{g}_{\mathbb{C}}) \subseteq \mathrm{End}(\mathfrak{g}_{\mathbb{C}})$. Since $\mathfrak{g}_{\mathbb{C}}$ is semi-simple $\mathrm{ad}(\mathfrak{g}_{\mathbb{C}})$ is the Lie algebra of <u>derivations</u> of $\mathfrak{g}_{\mathbb{C}}$ (see e.g. S. Helgason [14, Chapter II, Proposition 6.4]), that is, $D \in \mathrm{End}(\mathfrak{g}_{\mathbb{C}})$ lies in $\mathrm{ad}(\mathfrak{g}_{\mathbb{C}})$ iff

$$D[x,y] = [Dx,y] + [x,Dy] \ , \quad x,y \in \mathfrak{g}_{\mathbb{C}}.$$

We must show that if D is a derivation then also the semi-simple part is a derivation. So let $D = S + N$ be the Jordan decomposition. Then there is a direct sum decomposition $\mathfrak{g}_{\mathbb{C}} = \bigoplus_{\lambda} (\mathfrak{g}_{\mathbb{C}})_{\lambda}$ such that $(\mathfrak{g}_{\mathbb{C}})_{\lambda}$ is the eigenspace of S with eigenvalue λ, that is

$$\mathfrak{g}_{\mathbb{C}\lambda} = \{v \in \mathfrak{g}_{\mathbb{C}} \mid (D-\lambda)^k v = 0 \quad \text{for some} \quad k > 0\}.$$

That S is a derivation simply means that for $\lambda, \mu \in \mathbb{C}$,

$$[\mathfrak{g}_{\mathbb{C}\lambda}, \mathfrak{g}_{\mathbb{C}\mu}] \subseteq \mathfrak{g}_{\mathbb{C}(\lambda+\mu)}.$$

This, however, easily follows from the identity

$$(D-\lambda-\mu)^k[x,y] = \sum_{i=0}^{k} \binom{k}{i} [(D-\lambda)^{k-i}x, (D-\mu)^i y], \quad x,y \in \mathfrak{g}_{\mathbb{C}}, \quad k=0,1,2,\ldots,$$

which is proved by induction on k. This proves the lemma.

Now let $T \subseteq G$ be a maximal torus with Lie algebra $\mathfrak{t}$, let $\mathfrak{t}_{\mathbb{C}} = \mathfrak{t} \otimes_{\mathbb{R}} \mathbb{C} \subseteq \mathfrak{g}_{\mathbb{C}}$ and let $T_{\mathbb{C}} \subseteq G_{\mathbb{C}}$ be the corresponding connected Lie group. Every element $t \in \mathfrak{t}$ is semi-simple since $\mathrm{ad}(t) : \mathfrak{g} \to \mathfrak{g}$ is skew-adjoint with respect to a G-invariant metric. Therefore every element of $\mathfrak{t}_{\mathbb{C}}$ is semi-simple as well and we have the <u>root space decomposition</u> (see e.g. Helgason [14, Chapter III, § 4])

$$\mathfrak{g}_{\mathbb{C}} = \mathfrak{t}_{\mathbb{C}} \oplus \bigoplus_{\alpha \in \Phi} \mathfrak{g}_{\mathbb{C}\alpha},$$

where $\alpha : \mathfrak{t}_{\mathbb{C}} \to \mathbb{C}$, $\alpha \in \Phi$, are the roots, i.e. $\mathfrak{g}_{\mathbb{C}\alpha}$ are one-dimensional subspaces and

$$[t,x_\alpha] = \alpha(t) \cdot x_\alpha, \quad t \in \mathfrak{t}_{\mathbb{C}}, \; x_\alpha \in \mathfrak{g}_{\mathbb{C}\alpha}.$$

Furthermore let $\Phi^+ \subseteq \Phi$ be a choice of positive roots and let

(8.A.2) $$\mathfrak{b} = \mathfrak{t}_{\mathbb{C}} \oplus \bigoplus_{\alpha\in\Phi^+} \mathfrak{g}_{\mathbb{C}\alpha}, \quad \mathfrak{b}^+ = \bigoplus_{\alpha\in\Phi^+} \mathfrak{g}_{\mathbb{C}\alpha}.$$

Then both $\mathfrak{b}$ and $\mathfrak{b}^+$ are subalgebras of $\mathfrak{g}_{\mathbb{C}}$ since

(8.A.3) $$[\mathfrak{g}_{\mathbb{C}\alpha}, \mathfrak{g}_{\mathbb{C}\beta}] \subseteq \mathfrak{g}_{\mathbb{C}(\alpha+\beta)}, \qquad \alpha,\beta \in \Phi.$$

Also let $B \subseteq G_{\mathbb{C}}$ be the group with Lie algebra $\mathfrak{b}$. With this notation we now have

Lemma 8.A.4. a) $\mathfrak{t}_{\mathbb{C}}$ is a maximal abelian subalgebra of $\mathfrak{g}_{\mathbb{C}}$. Furthermore every element of $\mathfrak{t}_{\mathbb{C}}$ is semi-simple and every element of $\mathfrak{b}^+$ is nilpotent.

b) For every element $v \in \mathfrak{g}_{\mathbb{C}}$ there is $g \in G_{\mathbb{C}}$ such that $Ad(g)v = t+n \in \mathfrak{b}$ with $t \in \mathfrak{t}_{\mathbb{C}}$, $n \in \mathfrak{b}^+$ and $[t,n] = 0$. Furthermore, if $v \in \mathfrak{b}^+$, then the semi-simple part of v is conjugate to t.

c) The inclusion $NT \to NT_{\mathbb{C}}$ of normalizers of T and $T_{\mathbb{C}}$ in G and $G_{\mathbb{C}}$, respectively, induces an isomorphism

$$W = NT/T \xrightarrow{\cong} NT_{\mathbb{C}}/T_{\mathbb{C}}.$$

d) If $s \in \mathfrak{t}_{\mathbb{C}}$ and if for some $g \in G_{\mathbb{C}}$, $Ad(g)s \in \mathfrak{t}_{\mathbb{C}}$ then there exists $w \in NT_{\mathbb{C}}$ such that $Ad(w)s = Ad(g)s$.

Proof. a) For $v \in \mathfrak{g}_{\mathbb{C}}$ let $\bar{v}$ be the complex conjugate of v. If $[v,\mathfrak{t}_{\mathbb{C}}] = 0$ then clearly also $[\bar{v},\mathfrak{t}_{\mathbb{C}}] = 0$ so both the real and imaginary part $\operatorname{Re} v$ and $\operatorname{Im} v$ satisfy

$$[\operatorname{Re} v, \mathfrak{g}] = 0, \quad [\operatorname{Im} v, \mathfrak{g}] = 0$$

so by maximality of $\mathfrak{t}$ $v = \mathrm{Re}\, v + i\, \mathrm{Im}\, v = 0$. This shows that $\mathfrak{t}_{\mathbb{C}}$ is a maximal abelian subalgebra. The second statement is already proved and the last clearly follows from (8.A.3).

b) By the Iwasawa decomposition (see e.g. Helgason [14, Chapter VI, Theorem 6.3]) we have

(8.A.5) $$G_{\mathbb{C}} = G \cdot \exp(i\,\mathfrak{t}) \cdot \exp \mathfrak{b}^{+}$$

in particular $B \cap G = T$ and the inclusion $G \to G_{\mathbb{C}}$ induces a diffeomorphism

$$G/T \approx G_{\mathbb{C}}/B$$

so the Euler characteristic of $G_{\mathbb{C}}/B$ is different from zero (cf. Adams [1, proof of Theorem 4.21]). For $v \in \mathfrak{g}_{\mathbb{C}}$ we therefore conclude by Lefschetz' fixed point theorem that there is an element $g \in G_{\mathbb{C}}$ such that $gB \in G_{\mathbb{C}}/B$ is fixed under the one-parameter group of diffeomorphisms

$$h_r : G_{\mathbb{C}}/B \to G_{\mathbb{C}}/B, \quad r \in \mathbb{R},$$

where $h_r(xB) = \exp(rv)xB$, $r \in \mathbb{R}$, that is,

$$g^{-1}\exp(rv)g \in B, \quad \forall r \in \mathbb{R}.$$

Hence $\mathrm{Ad}(g^{-1})v \in \mathfrak{b}$. We can therefore suppose $v \in \mathfrak{b}$, and we write

(8.A.6) $$v = t + \sum_{\alpha \in \Phi^{+}} x_{\alpha}, \quad t \in \mathfrak{t}_{\mathbb{C}}, \quad x_{\alpha} \in \mathfrak{g}_{\mathbb{C}\alpha}.$$

Now we claim that we can change v by conjugation by elements of B so that $x_{\alpha} \neq 0$ only for $\alpha(t) = 0$. In fact suppose α is a minimal root so that both $x_{\alpha} \neq 0$ but $\alpha(t) \neq 0$. Then

$$\begin{aligned}\mathrm{Ad}(\exp(\tfrac{1}{\alpha(t)}\,x_\alpha))v &= \mathrm{Exp}(\mathrm{ad}(\tfrac{1}{\alpha(t)}\,x_\alpha))(v)\\ &= v - \frac{1}{\alpha(t)}[v,x_\alpha] + \sum_{i=2}^{\infty}\frac{1}{i!}(\mathrm{ad}(\tfrac{1}{\alpha(t)}\,x_\alpha))^i(v)\\ &= t + \sum_{\alpha'>\alpha} y_{\alpha'},\end{aligned}$$

where $\alpha' > \alpha$ means that $\alpha' - \alpha$ is a positive root. Iterating this procedure we can find $b \in B$ such that

$$\mathrm{Ad}(b)v = t + \sum_{\substack{\alpha\in\Phi^+\\ \alpha(t)=0}} z_\alpha .$$

Therefore we put $n = \sum_{\alpha\in\Phi^+} z_\alpha \in \mathfrak{b}^+$ and we clearly have $[t,n] = 0$; hence $\mathrm{Ad}(b)v = t + n$ is the Jordan decomposition. Notice that conjugation by $b \in B$ does not change the component in $\mathfrak{t}_{\mathbb{C}}$ in the decomposition (8.A.6) which proves the second statement in b).

c) Clearly $NT \subseteq NT_{\mathbb{C}}$ and since $T_{\mathbb{C}} \cap G = T$ the map $NT/T \to NT_{\mathbb{C}}/T_{\mathbb{C}}$ is injective. Now for $g \in T$ a regular element, left-multiplication by g

$$L_g : G_{\mathbb{C}}/B \to G_{\mathbb{C}}/B$$

has a fixed point for every element in $NT_{\mathbb{C}}/NT_{\mathbb{C}} \cap B$. Therefore the composite

$$NT/T \to NT_{\mathbb{C}}/T_{\mathbb{C}} \to NT_{\mathbb{C}}/\, NT_{\mathbb{C}} \cap B$$

is a bijection so it remains to show that $T_{\mathbb{C}} = NT_{\mathbb{C}} \cap B$. This, however, is trivial from the fact that every element of B is of the form $a \cdot \exp(n)$ with $a \in T_{\mathbb{C}}$ and $n \in \mathfrak{b}^+$. This ends the proof of c).

d) Let $s \in \mathfrak{t}_{\mathbb{C}}$ and $g \in G_{\mathbb{C}}$ with $\mathrm{Ad}(g)s = t \in \mathfrak{t}_{\mathbb{C}}$. Consider the Lie algebra

$$\mathfrak{s} = \{v \in \mathfrak{g}_{\mathbb{C}} \mid [v,t] = 0\}$$

and let $D \subseteq G_{\mathbb{C}}$ be the associated connected subgroup of $G_{\mathbb{C}}$. Then clearly $\mathfrak{t}_{\mathbb{C}} \subseteq \mathfrak{s}$ and also $\mathrm{Ad}(g)\,\mathfrak{t}_{\mathbb{C}} \subseteq \mathfrak{s}$ since for $x \in \mathfrak{t}_{\mathbb{C}}$

$$[\mathrm{Ad}(g)(x),t] = [x,s] = 0.$$

Also $\mathfrak{t}_{\mathbb{C}}$ and hence $\mathrm{Ad}(g)\,\mathfrak{t}_{\mathbb{C}}$ are Cartan subalgebras (i.e. a nilpotent algebra with itself as normalizer). Hence by the conjugacy theorem (see e.g. J. P. Serre [25, Chapitre III, Théorème 2]) there exists a $d \in D$ such that

$$\mathrm{Ad}(g)\,\mathfrak{t}_{\mathbb{C}} = \mathrm{Ad}(d)\,\mathfrak{t}_{\mathbb{C}}.$$

Hence $d^{-1}g \in NT_{\mathbb{C}}$ and $\mathrm{Ad}(d^{-1}g)s = \mathrm{Ad}(d)t = t$. This ends the proof of the lemma.

After these preparations we now return to the proof of the differentiability of $P' : \mathfrak{g} \to \mathbb{R}$ in the proof of Proposition 8.3. Recall that $\mathfrak{g}$ is the Lie algebra of a compact connected Lie group G with maximal torus T and P is a homogeneous polynomial of degree k on the Lie algebra $\mathfrak{t}$ of T. $P' : \mathfrak{g} \to \mathbb{R}$ is defined by the formula

$$P'(v) = P(\mathrm{ad}(g)v) \quad \text{where} \quad \mathrm{Ad}(g)v \in \mathfrak{t} \text{ for some } g \in G.$$

We shall show that P' extends to a complex analytic function $P'_{\mathbb{C}}$ on $\mathfrak{g}_{\mathbb{C}}$.

Since G is compact $\mathfrak{g} = \mathfrak{z} \oplus \mathfrak{g}'$ where

$$\mathfrak{z} = \{v \in \mathfrak{g} \mid [v,x] = 0 \quad \forall x \in \mathfrak{g}\}$$

is the center and $\mathfrak{g}'$ is a semi-simple ideal (see Helgason [14, Chapter II, Proposition 6.6]). Furthermore, if $Z \subseteq G$

is the center of G then $\mathfrak{g}'$ is naturally identified with the Lie algebra of the group $G' = G/Z$. Clearly the adjoint representation factors through G' and

$$Ad(g)(z+v) = z + Ad(g')v, \quad z \in \mathfrak{z}, \; v \in \mathfrak{g}', \; g \in G,$$

where $g' = gZ \in G'$. Also $T' = T/Z$ is a maximal torus in G' and $\mathfrak{t} = \mathfrak{z} \oplus \mathfrak{t} \cap \mathfrak{g}'$ where $\mathfrak{t} \cap \mathfrak{g}'$ is the Lie algebra of T'. Notice that G' is a compact semi-simple Lie group without center. Therefore we shall restrict to the case where G is semi-simple without center. The reader will have no difficulties in extending the arguments to the general case.

The homogeneous polynomial $P : \mathfrak{t} \to \mathbb{R}$ clearly extends to a <u>complex</u> homogeneous polynomial $P_{\mathbb{C}} : \mathfrak{t}_{\mathbb{C}} \to \mathbb{C}$ and obviously $P_{\mathbb{C}}$ is invariant under the adjoint action of $NT_{\mathbb{C}}$ by Lemma 8.A.4 c) and the invariance of P under the action by W on $\mathfrak{t}$. Now define $P'_C : \mathfrak{g}_{\mathbb{C}} \to \mathbb{C}$ as follows:

For $v \in \mathfrak{g}_{\mathbb{C}}$ choose $g \in G_{\mathbb{C}}$ such that

$$Ad(g)v = t + n$$

as in Lemma 8.A.4 b), and put

$$P'_{\mathbb{C}}(v) = P_{\mathbb{C}}(t).$$

Then this is clearly well-defined by the uniqueness of the Jordan-decomposition and Lemma 8.A.4 d). Clearly also $P'_{\mathbb{C}}|\mathfrak{g} = P'$.

First we show that $P'_{\mathbb{C}} : \mathfrak{g}_{\mathbb{C}} \to \mathbb{C}$ is <u>continuous</u>: For this let $\pi : \mathfrak{b} \to \mathfrak{t}_{\mathbb{C}}$ be the projection in the decomposition $\mathfrak{b} = \mathfrak{t}_{\mathbb{C}} \oplus \mathfrak{b}^+$ and notice that if $Ad(g)v = t + n$ as above then we can write $g = u \cdot b$, $u \in G$, $b \in B$ by (8.A.5) and then

$$Ad(u)v = Ad(b^{-1})(t+n) = t + n', \quad \text{with} \quad n' \in \mathfrak{b}^+.$$

It follows that

(8.A.7) $$P'_{\mathbb{C}}(v) = P_{\mathbb{C}}(\pi(\mathrm{Ad}(u)v))$$

and by the second part of Lemma 8.A.4 b) this equation holds for any $u \in G$ such that $\mathrm{Ad}(u)v \in \mathfrak{b}$.

To show that $P'_{\mathbb{C}}$ is continuous it suffices to show that whenever a sequence $\{v_k\}$, $k = 1,2,\ldots,$ converges to v, then there is a subsequence $\{v_{k_i}\}$ such that $P'_{\mathbb{C}}(v_{k_i}) \to P'_{\mathbb{C}}(v)$. Now choose $u_k \in G$ such that $\mathrm{Ad}(u_k)v_k \in \mathfrak{b}$. Since G is compact we can assume by taking a subsequence that u_k converges to u, say. Hence $\mathrm{Ad}(u_k)v_k \to \mathrm{Ad}(u)v$ and so

$$P'_{\mathbb{C}}(v_k) = P_{\mathbb{C}}(\pi(\mathrm{Ad}(u_k)v_k)) \to P_{\mathbb{C}}(\pi(\mathrm{Ad}(u)v)) = P'_{\mathbb{C}}(v).$$

To see that $P'_{\mathbb{C}}$ is actually <u>complex analytic</u> it suffices by the Riemann removable singularity theorem (cf. R. C. Gunning and H. Rossi [13, Chapter 1, § C, Theorem 3]) to show that it is complex analytic outside a closed algebraic set $S \subsetneqq \mathfrak{g}_{\mathbb{C}}$. For this consider the complex analytic mapping

$$F : G_{\mathbb{C}}/T_{\mathbb{C}} \times \mathfrak{t}_{\mathbb{C}} \to \mathfrak{g}_{\mathbb{C}}$$

defined by

$$F(g,t) = \mathrm{Ad}(g)t, \quad t \in \mathfrak{t}_{\mathbb{C}},\ g \in G_{\mathbb{C}},$$

and notice that $P'_{\mathbb{C}}(F(g,t)) = P_{\mathbb{C}}(t)$. It follows that $P'_{\mathbb{C}}$ is analytic near points $v = \mathrm{Ad}(g)t$ for which F is non-singular at (g,t). Now it is easy to see that F is singular at (g,t) only if t is <u>singular</u> in the sense that the kernel of $\mathrm{ad}(t) : \mathfrak{g}_{\mathbb{C}} \to \mathfrak{g}_{\mathbb{C}}$ is strictly bigger than $\mathfrak{t}_{\mathbb{C}}$. Now let $l = \dim_{\mathbb{C}} \mathfrak{t}_{\mathbb{C}}$ and let $S \subseteq \mathfrak{g}_{\mathbb{C}}$ be the set

$$S = \{v \in \mathfrak{g}_{\mathbb{C}} \mid \text{the semi-simple part } s \text{ of } v \text{ satisfies } \dim(\ker \mathrm{ad}(s)) > 1\} .$$

Notice that if $v \in \mathfrak{g}_{\mathbb{C}} - S$ then by Lemma 8.A.4 b), v is actually semi-simple so by the above $P'_{\mathbb{C}}$ is complex analytic near v. It remains to show that S is an algebraic subset different from $\mathfrak{g}_{\mathbb{C}}$: For this let

$$a_0(v) + a_1(v)\lambda + \ldots + a_n(n)\lambda^n = \det(\mathrm{ad}(v) - \lambda 1),\ n = \dim_{\mathbb{C}} \mathfrak{g}_{\mathbb{C}},$$

be the characteristic polynomial of $\mathrm{ad}\, v$. Then clearly

$$S = \{v \in \mathfrak{g}_{\mathbb{C}} \mid a_0(v) = \ldots = a_1(v) = 0\}$$

which is obviously a closed algebraic set and since

$$\mathfrak{t}_{\mathbb{C}} \cap S = \bigcup_{\alpha \in \Phi} \ker \alpha \neq \mathfrak{t}_{\mathbb{C}}$$

there exist elements outside S. This finishes the proof of the complex analyticity of $P'_{\mathbb{C}}$ and ends the proof of Proposition 8.3.

9. Applications to flat bundles

Again let G be an arbitrary Lie group with finitely many components. In Chapter 3 we called a connection in a principal differentiable G-bundle flat if the curvature form vanishes and we showed (Corollary 3.22) that this is equivalent to having a set of trivializations with constant transition functions, i.e. the bundle has a reduction to the group G_d, the underlying discrete group of G. This last condition of course also makes sense for topological G-bundles, so we shall take this as the definition of a flat G-bundle in general. Then by Theorem 5.5 the characteristic classes with coefficients in a ring Λ are in one-to-one correspondence with the elements of $H^*(BG_d,\Lambda)$. Let $j : G_d \to G$ be the natural map (actually the identity map) with corresponding map $Bj : BG_d \to BG$ of classifying spaces. The following proposition is obvious from Theorem 6.13 d):

Proposition 9.1. The following composite maps are zero

(i) $$I^*(G) \xrightarrow{w} H^*(BG,\mathbb{R}) \xrightarrow{Bj^*} H^*(BG_d,\mathbb{R}),$$

(ii) $$I^*_{\mathbb{C}}(G) \xrightarrow{w} H^*(BG,\mathbb{C}) \xrightarrow{Bj^*} H^*(BG_d,\mathbb{C}).$$

Corollary 9.2. a) The Chern classes with real coefficients of flat $Gl(n,\mathbb{C})$-bundles are zero.

b) The Pontrjagin classes with real coefficients of flat $Gl(n,\mathbb{R})$-bundles are zero.

From a differential geometric point of view these are just trivial remarks. However, a direct proof of Corollary 9.2 from the usual topological definitions of Chern classes is really not so easy. For this as well as for the general subject of this chapter see F. W. Kamber and Ph. Tondeur [16, especially Chapter 4] and also [16a]. (See also Exercise 3 below, and for a completely different point of view A. Grothendieck [12]).

Notice that if G is compact then by Theorem 8.1 we conclude that $Bj^*: H^*(BG,\mathbb{R}) \to H^*(BG_d,\mathbb{R})$ is zero. However, for G non-compact $w : I^*(G) \to H^*(BG,\mathbb{R})$ is in general not surjective and Bj^* need not be zero. For example J. Milnor has shown that there exist flat $Sl(2,\mathbb{R})$-bundles with non-zero Euler class (see J. Milnor [22], or Exercise 2 below). On the other hand we shall see that the image of Bj^* only depends on G/K where $K \subseteq G$ is a maximal compact subgroup. In the following we fix a choice of K. Since G_d is a discrete group $H^*(BG_d,\mathbb{R})$ has an explicit algebraic description. In fact for Π any discrete group the nerve $N\Pi$ is a discrete simplicial set and by Proposition 5.15, $H^*(B\Pi,\mathbb{R})$ is the homology of the complex $C^*(N\Pi)$ where a q-cochain is a function $c : \Pi \times \ldots \times \Pi \to \mathbb{R}$ (q factors of Π) and where the coboundary δ is given by

$$(9.3) \qquad \delta(c)(x_1,\ldots,x_{q+1}) = c(x_2,\ldots,x_{q+1}) + \\ + \sum_{i=1}^{q} (-1)^i c(x_1,\ldots,x_i x_{i+1},\ldots,x_{q+1}) + \\ + (-1)^{q+1} c(x_1,\ldots,x_q), \quad x_1,\ldots,x_{q+1} \in \Pi.$$

The homology of this complex is known as the Eilenberg-MacLane group cohomology of Π. In this chapter we shall study Bj^* by giving an explicit description of the composite map

$$(9.4) \qquad I^*(K) \xrightarrow{w} H^*(BK,\mathbb{R}) \underset{\cong}{\leftarrow} H^*(BG,\mathbb{R}) \xrightarrow{Bj^*} H^*(BG_d,\mathbb{R}) = H(C^*NG_d).$$

This is done in two steps:

Step I. By Chern-Weil theory $P \in I^l(K)$ defines a closed

G-invariant 2l-form on G/K.

Step II. Using the contractibility of G/K we define for any closed G-invariant q-form on G/K a q-cocycle in C^*NG_d.

Step I. Let $\mathfrak{g}$ and $\mathfrak{k}$ be the Lie algebras of G and K, respectively. Choose an inner product in $\mathfrak{g}$ which is invariant under the adjoint action of K, and let $\kappa : \mathfrak{g} \to \mathfrak{k}$ be the orthogonal projection onto $\mathfrak{k}$. By left-translation κ defines a 1-form $\theta_K \in A^1(G,\mathfrak{k})$ which clearly defines a connection in the principal K-bundle $G \to G/K$. Let Ω_K be the associated curvature form. Then by Chern-Weil theory $P \in I^l(K)$ defines a closed 2l-form $P(\Omega_K^l)$ on G/K. Notice that since θ_K by definition is invariant under the left G-action also Ω_K and hence $P(\Omega_K^l)$ are G-invariant, where again G acts on the left on G and G/K.

Step II. For this we introduce the following

Definition 9.5. A filling of G/K is a family of C^∞ singular simplices

$$\sigma(g_1,\ldots,g_p) : \Delta^p \to G/K, \quad g_1,\ldots,g_p \in G, \quad p = 0,1,2,\ldots$$

(so for $p = 0$ $\sigma(\emptyset) = o$ is some "base point", usually $o = \{K\}$) such that for $p = 1,2,\ldots,$

$$(9.6) \qquad \sigma(g_1,\ldots,g_p) \circ \varepsilon^i = \begin{cases} L_{g_1} \circ \sigma(g_2,\ldots,g_p), & i = 0, \\ \sigma(g_1,\ldots,g_i g_{i+1},\ldots,g_p), & 0 < i < p, \\ \sigma(g_1,\ldots,g_{p-1}), & i = p, \end{cases}$$

(Here $L_{g_1} : G/K \to G/K$ as usually is given by $L_{g_1}(gK) = g_1 gK$).

Lemma 9.7. There exist explicit fillings of G/K.

Proof. Let $o = \{K\} \in G/K$ be the base point and let $h_s : G/K \to G/K$, $s \in [0,1]$ be a C^∞ contraction of G/K to o, that is, $h_0(x) = o$ $\forall x \in G/K$ and $h_1 = \mathrm{id}$ (this can be explicitly constructed using the exponential map, cf. the reference given in the remark following Theorem 8.1). We can assume that h_s is constantly equal to o for s near zero by replacing h_s by $h_{\delta(s)}$, where $\delta : [0,1] \to [0,1]$ is a non-decreasing C^∞ function with $\delta(1) = 1$ and $\delta(s) = 0$ for s near zero.

Now we define $\sigma(g_1,\ldots,g_p)$ inductively as follows: For $p = 0$ $\sigma(\emptyset) = o$ and for $p = 1$ $\sigma(g_1) : \Delta^1 \to G/K$ is given by

$$\sigma(g_1)(t_0,t_1) = h_{t_1}(g_1 o).$$

For $p > 1$ consider Δ^p as the cone on the face spanned by $\{e_1,\ldots,e_p\} \subseteq \mathbb{R}^{p+1}$. Then the restriction of $\sigma(g_1,\ldots,g_p)$ to that face must be given by $L_{g_1} \circ \sigma(g_2,\ldots,g_p)$, and we extend this map to the cone using the contraction h_s. Explicitly

$$(9.8) \qquad \sigma(g_1,\ldots,g_p)(t_0,\ldots,t_p) =$$

$$= h_{1-t_0}[g_1\sigma(g_2,\ldots,g_p)(t_1/(1-t_0),\ldots,t_p/(1-t_0))].$$

It is now straightforward to check (9.6) inductively.

The merit of a filling σ of G/K is that it enables us to construct explicit Eilenberg-MacLane cochains: Consider the subcomplex $\mathrm{Inv}_G(A^*(G/K))$ of the de Rham complex $A^*(G/K)$ consisting of G-invariant forms (where the G-action is induced by the left G-action on G/K). Define the map

$$J : \mathrm{Inv}_G(A^*(G/K)) \to C^*(NG_d)$$

by

(9.9) $$J(\omega)(g_1,\ldots,g_p) = \int_{\Delta^p} \sigma(g_1,\ldots,g_p)^*\omega,$$

$$g_1,\ldots,g_p \in G,\ \omega \in A^p(G/K),\ p = 0,1,2,\ldots$$

Proposition 9.10. a) J is a chain map.

b) The induced map on homology

$$J_* : H(\mathrm{Inv}_G A^*(G/K)) \to H(C^*(NG_d)) = H^*(BG_d,\mathbb{R})$$

is independent of the choice of filling.

Proof. a) By Stoke's theorem and (9.6)

$$\begin{aligned} J(d\omega)(g_1,\ldots,g_{p+1}) &= \int_{\Delta^{p+1}} \sigma(g_1,\ldots,g_{p+1})^* d\omega \\ &= \int_{\Delta^p} [L_{g_1} \circ \sigma(g_2,\ldots,g_{p+1})]^*\omega + \\ &\quad + \sum_{i=1}^{p} (-1)^i \int_{\Delta^p} \sigma(g_1,\ldots,g_i g_{i+1},\ldots,g_{p+1})^*\omega + \\ &\quad + (-1)^{p+1} \int_{\Delta^p} \sigma(g_1,\ldots,g_p)^*\omega \\ &= \delta(J(\omega))(g_1,\ldots,g_{p+1}) \end{aligned}$$

since $L_{g_1}^*\omega = \omega$.

b) We give an alternative description of J_*: Consider the map of simplicial manifolds

$$\tilde{\gamma} : N(G_d;G/K) \to NG_d$$

where

$$N(G_d;G/K)(p) = NG_d(p) \times G/K$$

and the face operators are given by

$$\varepsilon_i(g_1,\ldots,g_p,gK) = \begin{cases} (g_2,\ldots,g_p,gK), & i = 0, \\ (g_1,\ldots,g_i g_{i+1},\ldots,g_p,gK), & 0 < i < p, \\ (g_1,\ldots,g_{p-1},g_p gK), & i = p. \end{cases}$$

$\tilde{\gamma}$ is just given by the projection onto the first factor. (Cf. the proof of Theorem 8.1. The realization of $\tilde{\gamma}$ is the fibre bundle with fibre G/K associated to $\gamma_{G_d} : EG_d \to BG_d$. Notice that if σ is a filling of G/K then the family

$$L_{(g_1 \cdots g_p)^{-1}} \circ \sigma(g_1,\ldots,g_p) : \Delta^p \to G/K,\ g_1,\ldots,g_p \in G,\ p = 0,1,2,\ldots,$$

defines a section of $\|\tilde{\gamma}\|$ which explains the definition).

Now if $\omega \in A^q(G/K)$ is an invariant form then the corresponding family of forms on $\Delta^p \times NG_d(p) \times G/K$, $p = 0,1,\ldots,$ induced by the projections onto G/K, defines an element $\bar{\omega} \in A^q(N(G_d;G/K))$. Clearly $d\bar{\omega} = \overline{d\omega}$, so we have an induced map on homology $\bar{\ } : H(\mathrm{Inv}_G\, A^*(G/K)) \to H(A^*(N(G_d;G/K)))$. On the other hand, since G/K is contractible

$$\tilde{\gamma} : N(G_d;G/K) \to NG_d$$

induces an isomorphism in de Rham cohomology by Lemma 5.16 and Theorem 6.4. Hence the composite map

$$\begin{array}{ccc} H(\mathrm{Inv}_G\, A^*(G/K)) & \xrightarrow{\ \bar{\ }\ } & H(A^*(N(G_d;G/K))) & \\ & & \cong \Big\uparrow \tilde{\gamma}^* & \\ & & H(A^*(NG_d)) & \xrightarrow{\ I\ } H(C^*NG_d) \end{array}$$

is canonically defined (i.e. without a choice of filling) and we claim that this is just J_* In fact given a filling σ we get an explicit inverse to $\tilde{\gamma}^*$

$$\tilde{\sigma}^* : A^*(N(G_d;G/K)) \to A^*(NG_d)$$

where $\tilde{\sigma} : \Delta^p \times NG_d(p) \to \Delta^p \times NG_d(p) \times G/K$, $p = 0,1,2,\ldots,$ is given by

$$\tilde{\sigma}(t,(g_0,\ldots,g_p)) = (t,(g_0,\ldots,g_p),(g_1\cdots g_p)^{-1}\sigma(g_1,\ldots,g_p)(t))$$

$$t \in \Delta^p,\quad g_1,\ldots,g_p \in G,\ p = 0,1,2,\ldots$$

Then obviously for $\omega \in \mathrm{Inv}_G A^p(G/K)$

$$I(\widetilde{\sigma}^*\overline{\omega})(g_1,\ldots,g_p) = \int_{\Delta^p}[L_{(g_1\cdots g_p)^{-1}} \circ \sigma(g_1,\ldots,g_p)]^*\omega$$
$$= \int_{\Delta^p}\sigma(g_1,\ldots,g_p)^*\omega = J(\omega)(g_1,\ldots,g_p).$$

This proves the proposition.

Remark. In the proof of Lemma 9.7 we replaced the contraction h_s by the contraction $h_{\delta(s)}$ where $\delta(s) = 0$ for s near zero in order to be able to define the C^∞ map $\sigma(g_1,\ldots,g_p)$ on all of Δ^p. On the other hand the inductive construction (9.8) using the original contraction makes sense on the open simplex and the corresponding change of parameter does not affect the value of the integral (9.9). In particular let us describe h_s explicitly for the case where G is semisimple with finite center: Then we can choose a Cartan decomposition $\mathfrak{g} = \mathfrak{k} \oplus \mathfrak{p}$ (see Helgason [14, Chapter 3, § 7]) and the map $\varphi = \pi \circ \exp : \mathfrak{p} \to G/K$ (where $\pi : G \to G/K$ is the projection and $\exp : \mathfrak{g} \to G$ the exponential map) is a diffeomorphism (see Helgason [14, Chapter 6, Theorem 1.1]). Therefore we get a contraction defined by

(9.11) $$h_s(x) = \varphi(s\varphi^{-1}(x)), \quad x \in G/K,\ s \in [0,1].$$

The curves $s \mapsto h_s(x)$ are geodesics with respect to a G-invariant Riemannian metric on G/K and we shall therefore refer to the corresponding filling defined inductively by (9.8) as the filling by geodesic simplices.

We can now describe the composite map (9.4):

Theorem 9.12. For $P \in I^1(K)$ the image under $Bj^* : H^*(BG,\mathbb{R}) \to H^*(BG_d,\mathbb{R})$ of $w(P) \in H^{21}(BK,\mathbb{R}) \cong H^{21}(BG,\mathbb{R})$ is represented in $H^{21}(C^*(NG_d))$ by the Eilenberg-MacLane cochain

$J(P(\Omega_K^1))$, where $P(\Omega_K^1) \in \mathrm{Inv}_G(A^{2l}(G/K))$ is defined in step I above and J is given by (9.9). That is,

$$(9.13) \qquad Bj^*(\omega(P))(g_1,\ldots,g_{2l}) = \int_{\Delta^{2l}} \sigma(g_1,\ldots,g_{2l})^* P(\Omega_K^1)$$

where σ is a filling of G/K.

Proof. Let $i : K \hookrightarrow G$ be the inclusion and consider the commutative diagram of simplicial manifolds

$$(9.14) \qquad \begin{array}{ccccc} N(G_d;G/K) & \xrightarrow{\tilde{j}} & N\bar{G}/K & \xleftarrow{N\bar{i}} & N\bar{K}/K \\ \downarrow \tilde{\gamma} & & \downarrow & & \downarrow \cong \\ N(G_d) & \xrightarrow{Nj} & NG & \xleftarrow{Ni} & NK \end{array}$$

where $\tilde{j} : NG_d(p) \times G/K \to N\bar{G}(p)/K$ is given by

$$\tilde{j}(g_1,\ldots,g_p,gK) = (g_1\ldots g_p g,\ldots,g_p g, g)K.$$

In the diagram (9.14) all maps except $\tilde{j}$ and Nj induce isomorphisms in de Rham cohomology. Therefore we shall calculate

$$\tilde{j}^* \circ (N\bar{i})^{*-1} : H(A^*(NK)) \to H(A^*(N(G_d;G/K))).$$

For this let $\kappa : \mathfrak{g} \to \mathfrak{k}$ be the orthogonal projection as in step I and let θ be the canonical connection in $N\bar{G} \to NG$ given by (6.12). Then $\theta_K = \kappa \circ \theta$ is a connection in the principal K-bundle $N\bar{G} \to N\bar{G}/K$ and we let Ω_K be the curvature form. Notice that the restriction of θ_K and Ω_K to $N\bar{G}(0) = G$ are obviously the connection and curvature forms defined in step I above. For $P \in I^l(K)$, $(N\bar{i})^{*-1}\omega(P) \in H^{2l}(A^*(N\bar{G}/K))$ is clearly represented by the form $P(\Omega_K^1) \in A^{2l}(N\bar{G}/K)$. It follows that

$$\tilde{j}^* \circ (N\bar{i})^{*-1}\omega(P) \in H^{2l}(A^*(N(G_d;G/K)))$$

is represented by the form $\overline{P(\Omega_K^1)}$ where now $P(\Omega_K^1) \in A^{2l}(G/K)$

denotes the G-invariant form defined in step I and where $\bar{\omega} \in A^*(N(G_d;G/K))$ for $\omega \in \mathrm{Inv}_G A^*(G/K)$ is the associated simplicial form as in the proof of Proposition 9.10 b). Therefore it follows from the diagram (9.14) that $\overline{P(\Omega_K^1)}$ represents

$$\tilde{\gamma}^*(Nj)^*(Ni)^{*-1}(w(P)) \in H(A^*(N(G_d;G/K)))$$

and the theorem follows from the description of J_* given in the proof of Proposition 9.10 b).

As an example we shall now study Theorem 9.12 in the case $G = Sp(2n,\mathbb{R})$, the real <u>symplectic</u> group. This is the subgroup of non-singular matrices $g \in Gl(2n,\mathbb{R})$ such that ${}^t gJg = J$ where ${}^t g$ is the transpose of g and J is the matrix

$$J = \begin{pmatrix} 0 & 1 \\ -1 & 0 \end{pmatrix}.$$

Here the maximal compact subgroup is $K = G \cap O(2n)$ ($g \in O(2n)$ iff $g\,{}^t g = 1$) which is isomorphic to the unitary group $U(n)$ (equivalently $U(n) \subseteq Sp(2n,\mathbb{R})$ is the subgroup of elements commuting with J). The first class to study is therefore the first Chern-class $c_1 \in H^2(BU(n),\mathbb{R})$. First some notation:

Let $P(2n,\mathbb{R}) \subseteq Gl(2n,\mathbb{R})$ be the set of positive definite symmetric matrices. Let $M(2n,\mathbb{R})$ be the set of all $2n \times 2n$ matrices and $S(2n,\mathbb{R}) \subseteq M(2n,\mathbb{R})$ the set of symmetric matrices. Then the exponential map $\exp : S(2n,\mathbb{R}) \to P(2n,\mathbb{R})$ is a diffeomorphism with inverse log. We then have

<u>Theorem 9.15</u>. The image $(Bj)^*c_1 \in H^2(BSp(2n,\mathbb{R})_d;\mathbb{R})$ of the first Chern class is represented by the cochain

$$(9.16) \qquad (Bj^*c_1)(g_1,g_2) = -\frac{1}{4\pi}\int_0^1 \mathrm{tr}(J[{}^t g_1 g_1 + (g_2\,{}^t g_2)^{-s}]^{-1} \log g_2\,{}^t g_2)\,ds$$

where tr means trace.

Remark. Notice that ${}^t g_1 g_1 + ({g_2}^t g_2)^{-s}$ is a positive definite symmetric matrix hence invertible, so the right hand side is well-defined.

Before proving Theorem 9.15 let us specialize to the case $n = 1$. Then $G = Sp(2,\mathbb{R}) = Sl(2,\mathbb{R})$ the group of 2×2 matrices of determinant 1. Here $K = SO(2)$ and c_1 equals the Euler class $e \in H^2(BSO(2),\mathbb{R})$. For $g_1, g_2 \in G$ write

$$g_2{}^t g_2 = k^{-1}\begin{pmatrix} y & 0 \\ 0 & y^{-1} \end{pmatrix} k, \quad y > 0,\ k \in SO(2),$$

and

$$k^{-1}\,{}^t g_1 g_1 k = \begin{pmatrix} a & b \\ b & d \end{pmatrix}, \quad ad - b^2 = 1, \quad a,d > 0.$$

It is easy to see that (9.16) then reduces to

$$(Bj^*e)(g_1,g_2) = \frac{b}{2\pi}\int_0^1 \frac{\log y\, ds}{dy^{-s}+ay^{s}+2} = \frac{b}{2\pi}\int_0^y \frac{dt}{at^2+2t+d}$$

$$= \frac{1}{2\pi}\left[\operatorname{Arc\,tan}\left(\frac{1+ay}{b}\right) - \operatorname{Arc\,tan}\left(\frac{1+a}{b}\right)\right]$$

(and equal to zero for $b = 0$). Notice that the numerical value satisfies

(9.17) $$|(Bj^*e)(g_1,g_2)| < \frac{1}{2\pi}\cdot\frac{\pi}{2} = \frac{1}{4}.$$

(This inequality can also be deduced directly from Theorem 9.12; see Exercise 2 below). This has the following consequence due to J. Milnor [22]:

Corollary 9.18. Let $\xi : E \to X_h$ be a flat $Sl(2,\mathbb{R})$-bundle over an oriented surface X_h of genus $h > 1$. Then the Euler class $e(E)$ satisfies

(9.19) $$|\langle e(E),[X_h]\rangle| < h.$$

<u>Proof</u>. We first need some well-known facts about the topology of surfaces. X_h can be constructed as a 4h-polygon with pairwise identifications of the sides $x_i \sim x_i'$ as on the figure

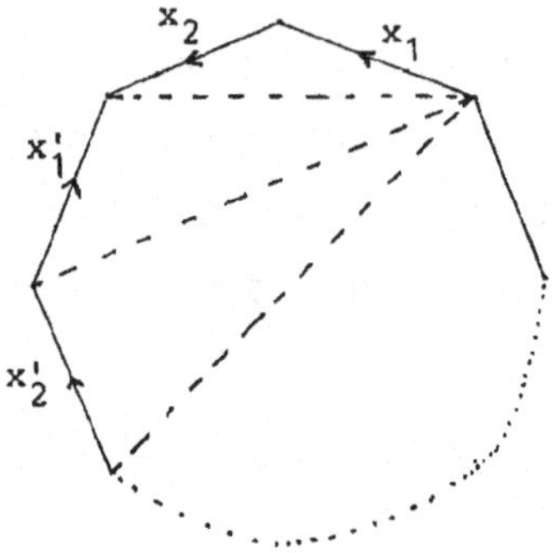

Here the sides $x_1,\ldots,x_{2h}$ give generators of the fundamental group Γ with the single relation

$$x_1 x_2 x_1^{-1} x_2^{-1} \ldots x_{2h-1}^{-1} x_{2h}^{-1} = 1 .$$

Furthermore the universal covering is contractible (see reference in Exercise 2 e) below). We can now define a continuous map $f : B\Gamma \to X_h$ as follows: For $x \in \Gamma$ choose a word in the generators $x_1,\ldots,x_{2h}$ representing x and map $\Delta^1 \times x \subseteq \Delta^1 \times \Gamma$ into the corresponding curve in the polygon. Now extend the map over the skeletons of $B\Gamma$ using the fact that the homotopy groups $\pi_i(X_h) = 0$ for $i > 1$. Clearly f is a homotopy equivalence by Whitehead's theorem. In particular the homology with integral coefficients of X_h is isomorphic to the homology of the complex $C_*N\Gamma$. Hence $H_2(C_*N\Gamma) \cong \mathbb{Z}$ and we claim that the generator is represented by the chain $z \in C_2(N\Gamma)$ defined by

$$z = (x_1,x_2) + (x_1x_2,x_1^{-1}) + \ldots + (x_1x_2x_1^{-1}x_2^{-1}\ldots x_{2h},x_{2h-1}^{-1}) +$$
$$+ (1,1) - (x_1,x_1^{-1}) + (1,1) - (x_2,x_2^{-1}) + \ldots + (1,1) - (x_{2h-1},x_{2h-1}^{-1})$$

which is easily checked to be a cycle. In fact $f_*z \in C_*(X_h)$ is the sum of all the $(4h-2)$ 2-simplices in the triangulation shown in the above figure plus some degenerate simplices.

Now any flat $Sl(2,\mathbb{R})$-bundle $\xi : E \to B\Gamma$ is induced by a map $B\alpha : B\Gamma \to BSl(2\mathbb{R})_d$ where $\alpha : \Gamma \to Sl(2,\mathbb{R})$ is a homomorphism (see Exercise 1 below). It follows that

$$\langle e(E), z\rangle = \langle Bj^*e, B\alpha_* z\rangle .$$

Now it is easy to see from (9.16) that a simplex of the form (x,x^{-1}) contribute zero (since in this case the integrand is the trace of the product of a skew-symmetric and a symmetric matrix). Therefore the right hand side consists of $4h-2$ terms each of which numerically contribute with less than 1/4. This proves the corollary.

<u>Proof of Theorem 9.15</u>. It is straightforward to check that $G = Sp(2n,\mathbb{R})$ is semi-simple so we can apply Theorem 9.12 using the filling σ by geodesic simplices. First let us reduce the number of integration variables:

In general for G semi-simple with maximal compact group K and Cartan decomposition $\mathfrak{g} = \mathfrak{k} \oplus \mathfrak{p}$ we have the diffeomorphism

$$\varphi = \pi \circ \exp : \mathfrak{p} \to G/K$$

as in the remark following Proposition 9.11. Therefore $\iota = \exp \circ \varphi^{-1} : G/K \to G$ is an embedding such that the diagram

$$\begin{array}{ccc} G/K & \xrightarrow{\ \iota\ } & G \\ & {\scriptstyle id}\searrow & \downarrow{\scriptstyle \pi} \\ & & G/K \end{array}$$

commutes. Then we have

<u>Lemma 9.20</u>. For $P \in I^1(K)$ and $g_1, g_2 \in G$,

$$(9.21)\qquad J(P(\Omega_K))(g_1,g_2) = \int_{\rho(g_1,g_2)} \iota^*P(\theta_K)$$

where $\rho(g_1,g_2)$ is the geodesic curve in G/K from g_1o to g_1g_2o (that is, $\rho(g_1,g_2)(s) = g_1\varphi(s\varphi^{-1}(g_2o))$, $s \in [0,1]$).

Proof. $P(\Omega_K)$ considered as a form on G is actually exact; in fact $P([\theta_K,\theta_K]) = 0$ since P is K-invariant, hence by (3.14)

$$P(\Omega_K) = d(P(\theta_K)) \quad \text{on} \quad G$$

and so

$$(9.22)\qquad P(\Omega_K) = d(\iota^*P(\theta_K)) \quad \text{on} \quad G/K.$$

Now by (9.8) the geodesic 2-simplex $\sigma(g_1,g_2) : \Delta^2 \to G/K$ is given by

$$(9.23)\qquad \sigma(g_1,g_2)(t_0,t_1,t_2) = h_{t_1+t_2}(g_1h_{t_2/(t_1+t_2)}(g_2o))$$

where $h_s(x) = \varphi(s\varphi^{-1}(x))$, $x \in G/K$, $s \in [0,1]$. Notice that θ_Y vanishes on the tangent fields along any curve of the form $\exp(sv)$, $s \in [0,1]$, and since $\iota \circ \sigma(g_1,g_2) \circ \varepsilon^i$, $i = 1,2$, is of this form we conclude from (9.22) that

$$\begin{aligned} J(P(\Omega_K))(g_1,g_2) &= \int_{\Delta^2} \sigma(g_1,g_2)^*d(\iota^*P(\theta_K)) \\ &= \int_{\Delta^1} (\sigma(g_1,g_2) \circ \varepsilon^0)^*\iota^*P(\theta_K) \end{aligned}$$

which is just (9.21).

Now for $G = Sp(2n,\mathbb{R}) \subseteq Gl(2n,\mathbb{R})$, the Lie algebra $\mathfrak{g} = \mathfrak{sp}(2n,\mathbb{R})$ is contained in $M(2n,\mathbb{R})$ as the set of matrices

$$\mathfrak{sp}(2n,\mathbb{R}) = \left\{X = \begin{pmatrix} A & B \\ C & -{}^tA \end{pmatrix} \middle|\, {}^tC = C,\ {}^tB = B\right\}.$$

The Lie algebra $\mathfrak{k} = \mathfrak{u}(n)$ of $K = U(n)$ is the subspace

$$\mathfrak{u}(n) = \left\{X = \begin{pmatrix} A & -C \\ C & A \end{pmatrix} \middle|\, {}^tC = C,\ {}^tA = -A\right\}$$

with complement in $\mathfrak{g}(2n,\mathbb{R})$:

$$\mathfrak{p} = \left\{A = \begin{pmatrix} A & B \\ B & -A \end{pmatrix} \middle| \ {}^tA = A, \quad {}^tB = B\right\}.$$

$\mathfrak{u}(n)$ is identified with the vectorspace of Hermitian $n \times n$ complex matrices (as in Example 5 of Chapter 4) by letting

$$X = \begin{pmatrix} A & -C \\ C & A \end{pmatrix}$$

correspond to $\hat{X} = A + iC$. In this notation the first Chern class $c_1 \in H^2(BU(n),\mathbb{R})$ is given by the Chern-Weil image of the linear form $P \in I^1(U(n))$ given by

$$(9.24) \qquad P(X) = -\frac{1}{2\pi i}\operatorname{tr}(\hat{X}) = -\frac{1}{2\pi}\operatorname{tr}(C) = -\frac{1}{4\pi}\operatorname{tr}(JX), \quad X \in \mathfrak{u}(n).$$

Now G/K is identified with $G \cap P(2n,\mathbb{R})$ via the map $\mu : G/K \to Gl(2n,\mathbb{R})$ given by

$$\mu(gK) = g\,{}^tg, \quad g \in G$$

(see G. Mostow [23, p. 20]). Under this identification the embedding $\iota : G/K \to G$ above is given by

$$\iota(p) = p^{\frac{1}{2}}, \quad p \in G \cap P(2n,\mathbb{R}).$$

Also if $\rho = \rho(s)$, $s \in [0,1]$, is a curve in $G \cap P(2n;\mathbb{R})$ let $\dot{\rho}$ denote the derivative, i.e. the tangent vector field along ρ.

Notice that the projection $\kappa : \mathfrak{g}(2n) \to \mathfrak{u}(n)$ is given by

$$\kappa(X) = \tfrac{1}{2}(X - {}^tX), \quad X \in \mathfrak{g}(2n).$$

For $P \in I^1(U(n))$ given by (9.24) above the form $\iota^*P(\theta_K)$ therefore takes the following form along a curve $\rho = \rho(s)$, $s \in [0,1]$, in $G \cap P(2n,\mathbb{R})$:

$$\iota^*P(\theta_K)(\dot\rho) = -\frac{1}{8\pi}\,\mathrm{tr}(J(\tau^{-1}\dot\tau - {}^t(\tau^{-1}\dot\tau))), \quad \tau = \rho^{\frac{1}{2}}.$$

But $\mathrm{tr}(J^t(\tau^{-1}\dot\tau)) = \mathrm{tr}(\tau^{-1}\dot\tau^t J) = -\,\mathrm{tr}(J\tau^{-1}\dot\tau)$ so

(9.25) $$\iota^*P(\theta_K)(\dot\rho) = -\frac{1}{4\pi}\,\mathrm{tr}(J\tau^{-1}\dot\tau), \quad \tau = \rho^{\frac{1}{2}}.$$

Now suppose ρ is a geodesic in $G \cap P(2n,\mathbb{R})$, that is,

$$\rho(s) = \tau_0 \exp(sY)\tau_0, \quad s \in [0,1],\ Y \in \mathcal{Y},$$
$$\tau_0 \in G \cap P(2n,\mathbb{R}).$$

Then

(9.26) $$\rho^{-1}\dot\rho = \tau_0^{-1}Y\tau_0 = \rho(0)^{-1}\dot\rho(0) = Q$$

is a constant in $\mathcal{Y}$. On the other hand, if we write $\rho(s) = \exp(Z(s))$, $Z(s) \in \mathcal{Y}$, $s \in [0,1]$, then (see Helgason [14, Chapter II, Theorem 1.7]):

$$\rho^{-1}\dot\rho = \frac{1-\exp(-\mathrm{ad}\ Z)}{\mathrm{ad}\ Z}(\dot Z)$$

$$= (1 + \exp(-\mathrm{ad}\ \tfrac{Z}{2}))\left(\frac{1-\exp(-\mathrm{ad}\ \frac{Z}{2})}{\mathrm{ad}\frac{Z}{2}}\right)(\dot{\tfrac{Z}{2}})$$

$$= (1 + \exp(-\mathrm{ad}\ \tfrac{Z}{2}))(\tau^{-1}\dot\tau),$$

where again $\tau = \rho^{\frac{1}{2}} = \exp\frac{Z}{2}$. Hence by (9.26)

$$\mathrm{tr}(J\tau^{-1}\dot\tau) = \mathrm{tr}(J(1 + \exp(-\mathrm{ad}\ \tfrac{Z}{2}))^{-1}(Q)).$$

Now since $Z \in S(2n,\mathbb{R})$, $\mathrm{ad}\,Z$ is a self adjoint transformation of $M(2n,\mathbb{R})$ with respect to the inner product

$$\langle A,B\rangle = \mathrm{tr}({}^tAB) = \mathrm{tr}(A^tB).$$

Therefore

$$\operatorname{tr}(J\tau^{-1}\dot{\tau}) = -\langle J, (1+\exp(-\operatorname{ad}\tfrac{Z}{2}))^{-1}(Q)\rangle$$
$$= \langle (1+\exp(-\operatorname{ad}\tfrac{Z}{2}))^{-1}(J), Q\rangle .$$

Now it is easy to see that $(\operatorname{ad}\frac{Z}{2})^k(J) = Z^k J$; hence

$$\operatorname{tr}(J\tau^{-1}\dot{\tau})' = -\langle (1+\exp(-Z))^{-1}J, Q\rangle$$
$$= \operatorname{tr}(J(1+\exp(-Z))^{-1}Q)$$
$$= \operatorname{tr}(J(1+\rho^{-1})^{-1}\rho(0)^{-1}\dot{\rho}(0)) .$$

Finally let $\rho = \rho(s)$, $s \in [0,1]$, be the geodesic curve from $g_1 o = g_1{}^t g_1$ to $g_1 g_2 o = g_1 g_2{}^t g_2{}^t g_1$, that is,

$$\rho(s) = g_1(g_2{}^t g_2)^s\,{}^t g_1, \quad s \in [0,1].$$

Then $\dot{\rho}(0) = g_1 \log(g_2{}^t g_2)$ and we conclude

$$\operatorname{tr}(J\tau^{-1}\dot{\tau})(s) = \operatorname{tr}(J[1+{}^t g_1^{-1}(g_2{}^t g_2)^{-s} g_1^{-1}]^{-1}\,{}^t g_1^{-1} \log(g_2{}^t g_2)\,{}^t g_1)$$
$$= \operatorname{tr}(J g_1^{-1}[1+{}^t g_1^{-1}(g_2{}^t g_2)^{-s} g_1^{-1}]^{-1}\,{}^t g_1^{-1} \log(g_2{}^t g_2))$$
$$= \operatorname{tr}(J[{}^t g_1 g_1 + (g_2{}^t g_2)^{-s}]^{-1} \log(g_2{}^t g_2))$$

since ${}^t g_1 J = J g_1^{-1}$. Theorem 9.15 now clearly follows from Theorem 9.12 together with (9.21) and (9.25).

Remark. It would be interesting to know if the expression in (9.16) is bounded also for $n > 1$.

Exercise 1. Let X be a connected locally path-connected and semi-locally 1-connected topological space so that it has a universal covering space $\pi : \tilde{X} \to X$. Let Γ be the fundamental group of X and let G be any Lie group.

a) Suppose $\alpha : \Gamma \to G$ is a homomorphism. Show that $\pi : \tilde{X} \to X$ is a principal Γ-bundle (therefore called a principal Γ-covering) and that the associated extension to a principal G-

bundle $\pi_\alpha : E_\alpha \to X$ is a flat G-bundle.

b) Suppose $\Gamma = \{1\}$ so that $X = \tilde{X}$ is simply connected. Show that every flat G-bundle is trivial. (Hint: Observe that the corresponding G_d-bundle is a covering space of X).

c) Show that in general every flat G-bundle on X is the extension of $\pi : \tilde{X} \to X$ to G relative to some homomorphism $\alpha : \Gamma \to G$.

Exercise 2. Let G be a Lie group with finitely many components and let $\alpha : \Gamma \to G$ be a homomorphism from a discrete group. Let $K \subseteq G$ be a maximal compact subgroup.

For $\omega \in \mathrm{Inv}_G A^*(G/K)$, the element $J_*\omega \in H^*(BG_d, \mathbb{R})$, defines a characteristic class for flat G-bundles.

a) Let $\pi : \tilde{M} \to M$ be a differentiable principal Γ-covering and let $\pi_\alpha : E_\alpha \to M$ be the corresponding flat G-bundle (see Exercise 1a) and let $\bar{\pi}_\alpha : \tilde{M} \times_\Gamma G/K \to M$ be the associated fibre-bundle with fibre G/K. Show that $\bar{\pi}_\alpha$ induces an isomorphism in cohomology and that the pull-back $\bar{\pi}_\alpha^*(J_*(\omega)(E_\alpha)) \in H^*(\tilde{M} \times_\Gamma G/K, \mathbb{R})$ of the characteristic class $J_*(\omega) \in H^*(M, \mathbb{R})$ is represented in $A^*(\tilde{M} \times_\Gamma G/K)$ by the unique form whose lift to $\tilde{M} \times G/K$ is just ω pulled back under the projection $\tilde{M} \times G/K \to G/K$.

b) Now suppose $\alpha : \Gamma \to G$ is the inclusion of a discrete subgroup such that $\pi : G/K \to \Gamma\backslash G/K = M_\Gamma$ is the covering space of a manifold (this is actually the case provided Γ is discrete and torsion free). Again let $\pi_\alpha : E_\alpha \to M_\Gamma$ be the associated flat G-bundle (first change the left Γ-action on G/K to a right action by $xg = g^{-1}x$ for $x \in G/K$, $g \in \Gamma$). Show that $J_*(\omega)(E_\alpha) \in H^*(M_\Gamma, \mathbb{R})$ is represented in $A^*(M_\Gamma)$ by the unique form $\hat{\omega}$ whose lift to G/K is just ω. (Hint: Observe that the diagonal $G/K \to G/K \times G/K$ induces a section of the bundle $\bar{\pi}_\alpha : \Gamma \backslash (G/K \times G/K) \to M_\Gamma$).

c) Again consider G, Γ and K as in b) and show that for $P \in I^1(K)$, $w(P)(E_\alpha) \in H^{2l}(M_\Gamma,\mathbb{R})$ is represented in $A^{2l}(M_\Gamma)$ by the form $P(\Omega_K^l)^\wedge$ where Ω_K is the curvature form of the connection given in step I. (Hint: Either use b) or give a direct proof by observing that $\pi_\alpha : E_\alpha \to M_\Gamma$ is the extension to G of the principal K-bundle $\Gamma\backslash G \to \Gamma\backslash G/K$). In particular, for $\dim G/K = 2k$,

$$(9.27) \qquad \langle w(P)(E_\alpha),\ [M_\Gamma]\rangle = \int_{M_\Gamma} P(\Omega_K^k), \quad \text{for all} \quad P \in I^k(K).$$

d) Let $\alpha_1 : \Gamma_1 \to G$ and $\alpha_2 : \Gamma_2 \to G$ be homomorphisms where Γ_1 and Γ_2 are the fundamental groups of two $2k$-dimensional compact manifolds M_1 and M_2 and let $\pi_{\alpha_1} : E_{\alpha_1} \to M_1$ and $\pi_{\alpha_2} : E_{\alpha_2} \to M_2$ be the corresponding flat G-bundles. Show the Hirzebruch proportionality principle:

There is a real constant $c(\alpha_1,\alpha_2)$ such that

$$(9.28) \qquad \langle w(P)(E_{\alpha_1}),[M_1]\rangle = c(\alpha_1,\alpha_2)\langle w(P)(E_{\alpha_2}),[M_2]\rangle \quad \text{for all} \quad P \in I^k(K).$$

Furthermore, if Γ_1 and Γ_2 are discrete subgroups of G and $M_i = M_{\Gamma_i}$, $i = 1,2$, as in b) above then $c(\alpha_1,\alpha_2) = vol(M_{\Gamma_1})/vol(M_{\Gamma_2})$ where M_{Γ_i}, $i = 1,2$, are given the Riemannian metrics induced from a left invariant metric on G/K (which exists since $\mathfrak{g}$ has an inner product which is invariant under the adjoint action by K).

e) Now consider $G = PSl(2,\mathbb{R}) = Sl(2,\mathbb{R})/\{\pm 1\}$. G acts by isometries on the Poincaré upper halfplane

$$H = \{z = x + iy \in \mathbb{C} \mid y > 0\}$$

with Riemannian metric

$$\frac{1}{y^2}(dx \otimes dx + dy \otimes dy).$$

The action is given by

$$z \longmapsto (az+b)/(cz+d), \qquad z \in \mathbb{C}$$

for

$$g = \begin{pmatrix} a & b \\ c & d \end{pmatrix} \in Sl(2,\mathbb{R}).$$

The isotropy subgroup at i is $K = SO(2)/\{\pm 1\}$ so we identify G/K with H. Here the Lie algebras are

$$\mathfrak{g} = \mathfrak{sl}(2,\mathbb{R}) = \left\{\begin{pmatrix} a & b \\ c & -a \end{pmatrix} \mid a,b,c \in \mathbb{R}\right\}$$

$$\mathfrak{k} = \mathfrak{o}(2) = \left\{\begin{pmatrix} 0 & -c \\ c & 0 \end{pmatrix} \mid c \in \mathbb{R}\right\}.$$

Let $\kappa : \mathfrak{g} \to \mathfrak{k}$ be the projection $\kappa(X) = \frac{1}{2}(X - {}^tX)$, $X \in \mathfrak{sl}(2,\mathbb{R})$, and let $P \in I^1(K)$ be the polynomial such that $\nu^*P = Pf$ where $\nu : SO(2) \to K$ is the projection and $Pf \in I^1(SO(2))$ is the Pfaffian.

i) Show that

$$(9.29) \qquad P(\Omega_K) = \frac{1}{4\pi}\, v$$

where v is the volume form on H.

It is well-known from non-Euclidean geometry (see e.g. C.L. Siegel [27, Chapter 3]) that there exist discrete subgroups $\Gamma \subsetneqq G$ acting discontinuously on H with quotient $\Gamma\backslash H$ a surface of genus, say h. In fact the fundamental domain of Γ is a non-Euclidean polygon with $4h$ sides.

ii) Check using the fact that the area of a non-Euclidean triangle ΔABC is $\pi - \angle A - \angle B - \angle C$, that the Euler characteristic of $\Gamma\backslash H$ is

$$\chi(\Gamma\backslash H) = 2(1-h).$$

(Hint: Observe first that the principal $SO(2)$-tangent bundle of G/K is the extension to $SO(2)$ of the principal K-bundle

$G \to G/K$ relative to the adjoint representation of K on the subspace $\mathfrak{q} = \ker(\kappa) \subseteq \mathfrak{sl}(2,\mathbb{R})$).

iii) Show that the inequality (9.17) follows from (9.29).

iv) Let $\Gamma \subseteq G$ with $\Gamma \backslash H$ a surface of genus h as above and let $\alpha : \Gamma^* \hookrightarrow Sl(2,\mathbb{R})$ be the inclusion of the inverse image of $\Gamma \subseteq G$. Let $\pi_\alpha : E_\alpha \to \Gamma \backslash H$ be the associated flat $Sl(2,\mathbb{R})$ bundle. Show that

(9.30) $$\langle e(E_\alpha), [\Gamma \backslash H] \rangle = h - 1.$$

Exercise 3. In this exercise we shall make a refinement of Corollary 9.2 using the topological definition of Chern classes as obstruction classes (see N. Steenrod [30, § 41]).

In general let G be a Lie group and F a manifold with a differentiable left G-action $G \times F \to F$. For $q \geq 0$ define a q-filling of F to be a family of C^∞ singular simplices

$$\sigma(g_1,\ldots,g_p) : \Delta^p \to F, \quad g_1,\ldots,g_p \in G, \; p = 0,1,2,\ldots,q,$$

such that (9.6) is satisfied for $p \leq q$.

a) Show that q-fillings exist if F is (q-1)-connected and that two q-fillings are homotopic (in the obvious sense) if F is q-connected.

b) Now suppose F is (q-1)-connected with q-filling σ and let $\omega \in \mathrm{Inv}_G(A^q(F,\mathbb{C}))$ be a closed complex valued G-invariant form representing an integral class (i.e. a class in the image of the inclusion $H^q(F,\mathbb{Z}) \subset H^q(F,\mathbb{C})$). Define the cochain $s(\omega) \in C^q(NG_d, \mathbb{C}/\mathbb{Z})$ by

(9.31) $$s(\omega)(g_1,\ldots,g_q) = \int_{\Delta^q} \sigma(g_1,\ldots,g_q)^*\omega$$

and show

i) $s(\omega)$ is a cocycle, hence defines a class

$$\hat{s}(\omega) \in H^q(BG_d, \mathbb{C}/\mathbb{Z}).$$

ii) $\hat{s}(\omega)$ does not depend on the choice of q-filling or choice of ω in the de Rham cohomology class.

iii) Suppose $H^q(F,\mathbb{Z}) \cong \mathbb{Z}$ and that ω represents a generator. If $\beta : H^q(BG_d, \mathbb{C}/\mathbb{Z}) \to H^{q+1}(BG_d, \mathbb{Z})$ is the Bockstein homomorphism then $\beta(\hat{s}(\omega))$ is the obstruction to the existence of a section of the universal G_d-bundle with fibre F over the q+1-skeleton of BG_d.

c) Let $G = Gl(n,\mathbb{C})$. For $\gamma_G : EG \to BG$ the universal G-bundle the k-th Chern class $c_k \in H^{2k}(BG,\mathbb{Z})$ is the obstruction to the existence of a section of the associated fibre bundle with fibre $F = Gl(n,\mathbb{C})/Gl(k-1,\mathbb{C})$. In fact F is 2k-2-connected and $H^{2k-1}(F,\mathbb{Z}) = \mathbb{Z}$. Show that there is a closed complex valued form $\omega_k \in Inv_G(A^{2k-1}(F,\mathbb{C}))$ representing the image of the generator in the de Rham cohomology with complex coefficients. (Hint: Observe that $Gl(n,\mathbb{C})$ is the complexification of $U(n)$ and notice that any cohomology class of $H^*(U(n)/U(k-1),\mathbb{R})$ can be represented by a $U(n)$-invariant real valued form). Conclude that if $j : Gl(n,\mathbb{C})_d \to Gl(n,\mathbb{C})$ is the natural map then

$$(9.32) \qquad Bj^*c_k = \beta(\hat{s}(\omega_k))$$

where $\hat{s}(\omega_k) \in H^{2k-1}(BGl(n,\mathbb{C})_d, \mathbb{C}/\mathbb{Z})$ is given by (9.31). In particular Bj^*c_k maps to zero in $H^{2k}(BG_d,\mathbb{C})$ which proves Corollary 9.2. (The classes $\hat{s}(\omega_k)$ have been introduced and studied by J. Cheeger and J. Simons (to appear)).

REFERENCES

[1] J. F. Adams, Lectures on Lie groups, W. A. Benjamin, New York - Amsterdam, 1969.

[2] P. Baum and R. Bott, On the zeroes of meromorphic vector fields, in: Essays on Topology and Related Topics, pp. 29-47, ed. A. Haefliger and R. Narasimhan, Springer-Verlag, Berlin - Heidelberg - New York, 1970.

[3] A. Borel, Sur la cohomologie des espaces fibres principaux et des espaces homogènes de groupes de Lie compacts, Ann. of Math. 57 (1953), pp. 115-207.

[4] R. Bott, Lectures on characteristic classes and foliations, in: Lectures on Algebraic and Differential Topology, pp. 1-94 (Lecture Notes in Math. 279), Springer-Verlag, Berlin - Heidelberg - New York, 1972.

[5] R. Bott, On the Chern-Weil homomorphism and the continuous cohomology of Lie groups, Advances in Math. 11 (1973), pp. 289-303.

[6] N. Bourbaki, Groupes et algèbre de Lie, Chapitres IV-VI, (Act. Sci. Ind. 1337), Hermann, Paris, 1968.

[7] G. Bredon, Sheaf Theory, McGraw-Hill, New York - London, 1967.

[8] H. Cartan, La transgression dans un groupe de Lie et dans un espace fibré principal, in: Colloque de topologie (Espace fibrés), pp. 57-71, George Thone, Liège, 1950.

[9] S. S. Chern and J. Simons, Characteristic forms and geometric invariants, Ann. of Math. 99 (1974), pp. 48-69.

[10] A. Dold, Lectures on Algebraic Topology, (Grundlehren Math. Wissensch. 200), Springer-Verlag, Berlin - Heidelberg - New York, 1972.

[11] J. L. Dupont, Simplicial de Rham cohomology and characteristic classes of flat bundles, Topology 15 (1976), pp. 233-245.

[12] A. Grothendieck, Classes de Chern et représentations linéaires des groupes discrets, in: Dix exposés sur la cohomologie des schémas, exp. VIII, pp. 215-305, North Holland Publ. Co., Amsterdam, 1968.

[13] R. C. Gunning and H. Rossi, Analytic functions of several variables, Prentice-Hall, Englewood Cliffs, 1965.

[14] S. Helgason, Differential Geometry and Symmetric Spaces, Academic Press, New York - London, 1962.

[15] G. Hochschild, The Structure of Lie Groups, Holden-Day, San Francisco - London - Amsterdam, 1965.

[16] F. W. Kamber and Ph. Tondeur, Flat Manifolds, (Lecture Notes in Math. 67), Springer-Verlag, Berlin - Heidelberg - New York, 1968.

[16a] F.W. Kamber and Ph. Tondeur, Foliated Bundles and Characteristic Classes, Lecture Notes in Mathematics 493, Springer-Verlag, Berlin-Heidelberg-New York,1975.

[17] S. Kobayashi and K. Nomizu, Foundations of Differential Geometry, I-II, (Interscience Tracts in Pure and Applied Math. 15), Interscience Publ., New York - London - Sydney, 1969.

[18] S. MacLane, Homology, (Grundlehren Math. Wissensch. 114), Springer-Verlag, Berlin - Göttingen - Heidelberg, 1963.

[19] J. W. Milnor and J. Stasheff, Characteristic classes, Annals of Math. Studies 76, Princeton University Press, Princeton, 1974.

[20] J. W. Milnor, Construction of Universal bundles, II, Ann. of Math. 63 (1956), pp. 430-436.

[21] J. W. Milnor, Morse Theory, Annals of Math. Studies 51, Princeton University Press, Princeton, 1963.

[22] J. W. Milnor, On the existence of a connection with curvature zero, Comment. Math. Helv. 32 (1958), pp. 215-223.

[23] G. Mostow, Strong Rigidity of Locally Symmetric Spaces, (Annals of Math. Studies 78), Princeton University Press, Princeton, 1973.

[24] G. Segal, Classifying spaces and spectral sequences, Inst. Hautes Études Sci. Publ. Math. 34 (1968), pp. 105-112.

[25] J. P. Serre, Algèbre Semi-Simple Complexes, W. A. Benjamin, New York, 1966.

[26] H. Shulmann, On Characteristic Classes, Thesis, University of California, Berkeley, 1972.

[27] C. L. Siegel, Topics in Complex Function Theory II, Automorphic Functions and Abelian Integrals, (Interscience Tracts in Pure and Applied Math. 25), Interscience Publ., New York, 1971.

[28] M. Spivak, Differential Geometry I, Publish or Perish, Boston, 1970.

[30] N. Steenrod, The Topology of Fibre Bundles, (Princeton Math. Series 14), Princeton University Press, Princeton, 1951.

[31] D. Sullivan, Differential forms and the topology of manifolds, in: Manifolds - Tokyo, 1973, pp. 37-49, ed. A. Hattori, University of Tokyo Press, Tokyo, 1975.

[32] B. L. van der Waerden, Algebra I, (Grundlehren Math. Wissensch. 33), Springer-Verlag, Berlin - Göttingen - Heidelberg, 1960.

[33] F. W. Warner, Foundations of Differentiable Manifolds and Lie groups, Scott, Foresman and Co., Glenview, 1971.

[34] A. Weil, Sur les théorèmes de de Rham, Comment. Math. Helv. 26 (1952), pp. 119-145.

[35] H. Whitney, Geometric Integration Theory, (Princeton Math. Series 21), Princeton University Press, Princeton, 1957.

/LD

LIST OF SYMBOLS

SUBJECT INDEX

Vol. 551: Algebraic K-Theory, Evanston 1976. Proceedings. Edited by M. R. Stein. XI, 409 pages. 1976.

Vol. 552: C. G. Gibson, K. Wirthmüller, A. A. du Plessis and E. J. N. Looijenga. Topological Stability of Smooth Mappings. V, 155 pages. 1976.

Vol. 553: M. Petrich, Categories of Algebraic Systems. Vector and Projective Spaces, Semigroups, Rings and Lattices. VIII, 217 pages. 1976.

Vol. 554: J. D. H. Smith, Mal'cev Varieties. VIII, 158 pages. 1976.

Vol. 555: M. Ishida, The Genus Fields of Algebraic Number Fields. VII, 116 pages. 1976.

Vol. 556: Approximation Theory. Bonn 1976. Proceedings. Edited by R. Schaback and K. Scherer. VII, 466 pages. 1976.

Vol. 557: W. Iberkleid and T. Petrie, Smooth S^1 Manifolds. III, 163 pages. 1976.

Vol. 558: B. Weisfeiler, On Construction and Identification of Graphs. XIV, 237 pages. 1976.

Vol. 559: J.-P. Caubet, Le Mouvement Brownien Relativiste. IX, 212 pages. 1976.

Vol. 560: Combinatorial Mathematics, IV, Proceedings 1975. Edited by L. R. A. Casse and W. D. Wallis. VII, 249 pages. 1976.

Vol. 561: Function Theoretic Methods for Partial Differential Equations. Darmstadt 1976. Proceedings. Edited by V. E. Meister, N. Weck and W. L. Wendland. XVIII, 520 pages. 1976.

Vol. 562: R. W. Goodman, Nilpotent Lie Groups: Structure and Applications to Analysis. X, 210 pages. 1976.

Vol. 563: Séminaire de Théorie du Potentiel. Paris, No. 2. Proceedings 1975–1976. Edited by F. Hirsch and G. Mokobodzki. VI, 292 pages. 1976.

Vol. 564: Ordinary and Partial Differential Equations, Dundee 1976. Proceedings. Edited by W. N. Everitt and B. D. Sleeman. XVIII, 551 pages. 1976.

Vol. 565: Turbulence and Navier Stokes Equations. Proceedings 1975. Edited by R. Temam. IX, 194 pages. 1976.

Vol. 566: Empirical Distributions and Processes. Oberwolfach 1976. Proceedings. Edited by P. Gaenssler and P. Révész. VII, 146 pages. 1976.

Vol. 567: Séminaire Bourbaki vol. 1975/76. Exposés 471–488. IV, 303 pages. 1977.

Vol. 568: R. E. Gaines and J. L. Mawhin, Coincidence Degree, and Nonlinear Differential Equations. V, 262 pages. 1977.

Vol. 569: Cohomologie Etale SGA 4½. Séminaire de Géométrie Algébrique du Bois-Marie. Edité par P. Deligne. V, 312 pages. 1977.

Vol. 570: Differential Geometrical Methods in Mathematical Physics, Bonn 1975. Proceedings. Edited by K. Bleuler and A. Reetz. VIII, 576 pages. 1977.

Vol. 571: Constructive Theory of Functions of Several Variables, Oberwolfach 1976. Proceedings. Edited by W. Schempp and K. Zeller. VI, 290 pages. 1977

Vol. 572: Sparse Matrix Techniques, Copenhagen 1976. Edited by V. A. Barker. V, 184 pages. 1977.

Vol. 573: Group Theory, Canberra 1975. Proceedings. Edited by R. A. Bryce, J. Cossey and M. F. Newman. VII, 146 pages. 1977.

Vol. 574: J. Moldestad, Computations in Higher Types. IV, 203 pages. 1977.

Vol. 575: K-Theory and Operator Algebras, Athens, Georgia 1975. Edited by B. B. Morrel and I. M. Singer. VI, 191 pages. 1977.

Vol. 576: V. S. Varadarajan, Harmonic Analysis on Real Reductive Groups. VI, 521 pages. 1977.

Vol. 577: J. P. May, E_∞ Ring Spaces and E_∞ Ring Spectra. IV, 268 pages. 1977.

Vol. 578: Séminaire Pierre Lelong (Analyse) Année 1975/76. Edité par P. Lelong. VI, 327 pages. 1977.

Vol. 579: Combinatoire et Représentation du Groupe Symétrique, Strasbourg 1976. Proceedings 1976. Edité par D. Foata. IV, 339 pages. 1977.

Vol. 580: C. Castaing and M. Valadier, Convex Analysis and Measurable Multifunctions. VIII, 278 pages. 1977.

Vol. 581: Séminaire de Probabilités XI, Université de Strasbourg. Proceedings 1975/1976. Edité par C. Dellacherie, P. A. Meyer et M. Weil. VI, 574 pages. 1977.

Vol. 582: J. M. G. Fell, Induced Representations and Banach *-Algebraic Bundles. IV, 349 pages. 1977.

Vol. 583: W. Hirsch, C. C. Pugh and M. Shub, Invariant Manifolds. IV, 149 pages. 1977.

Vol. 584: C. Brezinski, Accélération de la Convergence en Analyse Numérique. IV, 313 pages. 1977.

Vol. 585: T. A. Springer, Invariant Theory. VI, 112 pages. 1977.

Vol. 586: Séminaire d'Algèbre Paul Dubreil, Paris 1975–1976 (29ème Année). Edited by M. P. Malliavin. VI, 188 pages. 1977.

Vol. 587: Non-Commutative Harmonic Analysis. Proceedings 1976. Edited by J. Carmona and M. Vergne. IV, 240 pages. 1977.

Vol. 588: P. Molino, Théorie des G-Structures: Le Problème d'Equivalence. VI, 163 pages. 1977.

Vol. 589: Cohomologie l-adique et Fonctions L. Séminaire de Géométrie Algébrique du Bois-Marie 1965–66, SGA 5. Edité par L. Illusie. XII, 484 pages. 1977.

Vol. 590: H. Matsumoto, Analyse Harmonique dans les Systèmes de Tits Bornologiques de Type Affine. IV, 219 pages. 1977.

Vol. 591: G. A. Anderson, Surgery with Coefficients. VIII, 157 pages. 1977.

Vol. 592: D. Voigt, Induzierte Darstellungen in der Theorie der endlichen, algebraischen Gruppen. V, 413 Seiten. 1977.

Vol. 593: K. Barbey and H. König, Abstract Analytic Function Theory and Hardy Algebras. VIII, 260 pages. 1977.

Vol. 594: Singular Perturbations and Boundary Layer Theory, Lyon 1976. Edited by C. M. Brauner, B. Gay, and J. Mathieu. VIII, 539 pages. 1977.

Vol. 595: W. Hazod, Stetige Faltungshalbgruppen von Wahrscheinlichkeitsmaßen und erzeugende Distributionen. XIII, 157 Seiten. 1977.

Vol. 596: K. Deimling, Ordinary Differential Equations in Banach Spaces. VI, 137 pages. 1977.

Vol. 597: Geometry and Topology, Rio de Janeiro, July 1976. Proceedings. Edited by J. Palis and M. do Carmo. VI, 866 pages. 1977.

Vol. 598: J. Hoffmann-Jørgensen, T. M. Liggett et J. Neveu, Ecole d'Eté de Probabilités de Saint-Flour VI – 1976. Edité par P.-L. Hennequin. XII, 447 pages. 1977.

Vol. 599: Complex Analysis, Kentucky 1976. Proceedings. Edited by J. D. Buckholtz and T. J. Suffridge. X, 159 pages. 1977.

Vol. 600: W. Stoll, Value Distribution on Parabolic Spaces. VIII, 216 pages. 1977.

Vol. 601: Modular Functions of one Variable V, Bonn 1976. Proceedings. Edited by J.-P. Serre and D. B. Zagier. VI, 294 pages. 1977.

Vol. 602: J. P. Brezin, Harmonic Analysis on Compact Solvmanifolds. VIII, 179 pages. 1977.

Vol. 603: B. Moishezon, Complex Surfaces and Connected Sums of Complex Projective Planes. IV, 234 pages. 1977.

Vol. 604: Banach Spaces of Analytic Functions, Kent, Ohio 1976. Proceedings. Edited by J. Baker, C. Cleaver and Joseph Diestel. VI, 141 pages. 1977.

Vol. 605: Sario et al., Classification Theory of Riemannian Manifolds. XX, 498 pages. 1977.

Vol. 606: Mathematical Aspects of Finite Element Methods. Proceedings 1975. Edited by I. Galligani and E. Magenes. VI, 362 pages. 1977.

Vol. 607: M. Métivier, Reelle und Vektorwertige Quasimartingale und die Theorie der Stochastischen Integration. X, 310 Seiten. 1977.

Vol. 608: Bigard et al., Groupes et Anneaux Réticulés. XIV, 334 pages. 1977.